Asgeir Brekke · Alv Egeland

The Northern Light

From Mythology to Space Research

With 143 Black and White Photographs and Drawings
and 35 Coloured Photographs in 16 Plates

Springer-Verlag
Berlin Heidelberg New York Tokyo
1983

Dr. Asgeir Brekke
Institute of Mathematical and Physical Sciences
University of Tromsø
9001 Tromsø, Norway

Prof. Dr. Alv Egeland
Institute of Physics
University of Oslo
Blindern, Oslo 3, Norway

Title of the Norwegian Edition
A. Brekke og A. Egeland, Nordlyset
© 1979 by Grøndahl & Søn Forlag A.S., Oslo, Norway

ISBN-13: 978-3-642-69108-9   e-ISBN-13: 978-3-642-69106-5
DOI: 10.1007/978-3-642-69106-5

Library of Congress Cataloging in Publication Data. Brekke, Asgeir. The northern light. Translation of: Nordlyset. Bibliography: p.   Includes index. 1. Auroras. 2. Auroras–Scandinavia. I. Egeland, Alv, 1932– II. Title. QC971.B7313 1983   538′.768   83-8484

2132/3130-543210

Preface

In Nordic literature a remarkable discussion of the northern light appears in Kongespeilet (*The King's Mirror*) a thirteenth-century Norwegian chronicle. It is described in vivid detail as the following translated excerpts demonstrate:

> These northern lights have this peculiar nature, that the darker the night is, the brighter they seem, and they always appear at night but never by day, most frequently in the densest darkness and rarely by moonlight. In appearance they resemble a vast flame of fire viewed from a great distance. It also looks as if sharp points were shot from this flame up into the sky; these are of uneven height and in constant motion, now one, now another darting highest; and the light appears to blaze like a living flame.

Three different theories for the origin of the northern light were suggested in this book.

Numerous naturally occurring heavenly phenomena have been observed and enjoyed as long as the Earth has been inhabited, but hardly any of them has stirred man's imagination, curiosity and fear as much as the northern light. The northern light is certainly one of the most spectacular of nature's phenomena.

Unfortunately, this impressive and imaginative display on a dark, cold winter night has only been seen regularly by a limited number of people (i.e., those living near the polar regions). Consequently, historical records of the northern light are far less numerous than for other optical heavenly displays. Due to the very rare appearance of the northern light in densely populated areas of the world, records dating back more than a few hundred years are few and uncertain and mainly concern the morphological aspects.

Since Northern Scandinavia is located close to the zone of maximum auroral occurence, these majestic lights occurred so regularly that they were a routine part of the daily life of these northern people. It is therefore not surprising that the history of the northern light during the last 1000 years is better documented here than in any other area of the world.

The main purpose of this book is to review the historical contributions on the northern light by Danes, Norwegians and Swedes from the Viking period up to the first half of this century. The development of auroral history in the Nordic countries represents an interesting and fascinating story of the influence of the northern light on Scandinavian thought, up to our century. We also review some practical uses of the northern light in this part of the world, and discuss the impact of the northern light on Norwegian poetry and literature. In the latter part of the book emphasis is placed on the scientific contributions of Birkeland, Størmer and Vegard, who were the first to apply precise methods to the study of the aurora and to explain its light and associated magnetic disturbances as natural consequences of the arrival of fast electrons and ions at high latitudes. They were also the first to calculate the motion of such particles and to simulate the phenomenon in the laboratory. Through their research, these pioneers discovered many new effects and laid the foundation for our present-day exploration of the aurora from space. The subtitle of this book – *From mythology to space research* – reflects the broad nature of its contents.

This book was first published in Norwegian as *Nordlyset (The northern light)* in 1979 by Grøndahl & Søn, Oslo. However, this English version is significantly different from *Nordlyset*. Two chapters with special references to auroral institutions in Norway have been omitted, while the rest of the text has been updated and extended to broaden its scope.

The book is written in a form which requires no general background in mathematics or physics. Furthermore, it is not necessary (and even not recommended) to read this book through from beginning to end. It is really possible to read each chapter – and even individual sections – independently from the rest of the text.

It will be noted that the words "northern light" and "aurora" are used in this book to refer to the same phenomenon. "*Nordlyset*" in Norwegian translates into English as "The northern light", while the word aurora is of Latin origin, meaning "dawn".

In the Appendix a list of literature is presented for those who wish to study the northern light in more detail.

Acknowledgements

We have received much help and many suggestions from friends and colleagues in writing this book; many of them have generously allowed us to use their material and/or illustrations. Sam Silverman (Air Force Geophysics Lab., Mass., USA) allowed us to copy some articles from his extensive historical collection on the aurora. Practically all university libraries in Scandinavia have made their material available. In particular Ann-Grethe Holm-Olsen at Oslo University Library and Mrs. Gøril Hestvedt and Mr. Jens Lauesen at Tromsø University Library have been very helpful.

Heartfelt thanks are due to Steinar Berger (Auroral Observatory, Tromsø) for allowing us to use some of his excellent auroral photographs and to Chlarens Orsland for helping one of the authors (A.E.) with translating three chapters into English. It is also a pleasure to acknowledge the superb illustrations by Liv Larsen, The Auroral Observatory, Tromsø and photographic help by Terje Holm, Inst. of Phyiscs, University of Oslo.

Particular thanks are due to Grady Hicks (retired from Naval Research Laboratory) and his wife Violet, who during their last six months' stay in Norway helped us in translating the text from Norwegian to English. In addition Grady has read both the first and the second drafts of the manuscript and has offered most helpful advice and comments, practically on every page of the book, both in regard to English phraseology and scientific contents.

Finally, it is a pleasure to thank Mrs. Liv Larsen and Anne-Sophie Andresen for their patient and excellent work in typing the manuscript. Mrs. Andresen has also greatly helped in the practical work preparing this book.

We hope this book will stimulate and increase a general interest in this most spectacular of nature's phenomena – *The Northern Light*.

Summer 1983   ASGEIR BREKKE
                                                                                    ALV EGELAND

Contents

1 The Northern Light in Folklore and Mythology

1.1 The Enigmatic and Mysterious Northern Lights

From time Immemorial people have stopped in their tracks and become lost in thought when the northern lights have thrown their manifold flames into the sky. Fluttering draperies of innumerable spectral colours have often led the mind to wander into the realm of dancing spirits and fighting hordes.

Among the ancient Norwegians who were inspired by the northern lights, dance was no uncommon theme. In fact, inhabitants of the west coast of Norway even into the beginning of this century believed that the northern lights were "old maids" dancing and waving white gloved hands. In this part of the country it was believed that when old maids died they would be taken up to the northern lights. "She is so old that she soon will pass away to the northern lights" was a saying on the west coast of Norway, when old maids were spoken of unflatteringly.

Connections between old women and the northern light were common in Finland, as is evident in the following quotations referring to the northern lights and translated from the Finnish: "The women of the North are hovering in the air" or "The old women from Pohjanmaa hover at Konnunsuo", Konnunsuo being the place occupied by maidens after their death. It appears to have been a fairly common belief in the Nordic countries that old maids had some connection with the northern light. In Kangasniemi, Finland, the northern lights were simply referred to in the words: "The old maids are making fire".

The Lapps in Sweden even recite a jingle about girls and unmarried women when they sight a northern light. It goes as follows:

Neit, neit varret bira arnieb
månkan bele göseje

which translated freely into English is:

Girls, girls running around the fireplace
dragging their pants.

Many scholars of the last century claimed that in Norwegian mythology during the Viking period, the northern lights have been referred to as reflections from the shields of the Valkyrjes. The Valkyries were in fact dead maidens, and therefore this association of the old women and the northern lights might have been a very old one. The northern lights and their connection to Norse mythology will be discussed more thoroughly in Chap. 2.

The well-known Scottish expression for the northern lights, the Merry Dancers, is another example of the northern lights being associated with women. According to the legend these Merry Dancers were supernatural creatures fighting in the sky for the favour of a beautiful lady.

North American Indians also usually described the northern lights in terms of merry dancers, but for these Indians the northern light aroused no erotic visions as they did among the Scots. In the Indian's imagination, the northern lights were the gods dancing across the firmament. To the Swedes, the northern lights were also often related to dance and a plain old name for the phenomenon in Sweden is Polka. (Polka is a well-known folkdance.)

Among the Eskimos in Greenland and the Hudson Bay area, the northern light was the realm of the dead. The Greenlander Eskimos believed that a flickering northern light signified that their dead friends were trying to contact surviving relatives. The Greenlanders respected the northern light deeply and therefore avoided making fun of it. They believed that by whistling to the northern light they could accelerate its motion, and to perceive a rustling sound meant that contact had been established with their dead friends.

The Fox Indians believed they could conjure up ghosts and spirits by whistling to the northern lights. The approach of these spirits was said to sound like the pit-a-pat of bare feet on hard ground. They came running, bodies leaning forward, arms extended backward, glancing wildly this way and that; soaring through the air, they landed on the ground with a

1

Naturally enough, in older times it was a fairly world-wide belief that the northern lights were some kind of fire. Such examples can also be found among our ancestors in the Nordic countries, who believed that the northern lights were active volcanoes in the far north placed there by the mighty God himself to light up and heat the dark and cold parts of the country. The world-famous scientist Anders Celsius (cf. Chap. 5.8) even noted in his diary on September 24th 1732 that the northern light was caused by active volcanoes close to the North Pole.

In Sweden the belief also existed that the northern light was due to reflections from torches which the Lapps used when they moved about looking for their reindeer. In Finland, angels fighting, bearing flaming torches, were thought to be the cause of the northern light.

It is only natural that any tribe or group of people would seek an understanding of this unique phenomenon in terms of familiar events of their immediate neighbourhood and everyday life. With the Ostyaks, for instance, for whom fishing was daily work, the northern light was due to flames supplied by the god of fishing to help the fishermen who were out working late at night.

In Norway and Sweden it was a common belief that the northern lights were reflections from huge herring runs or of fish schools in the ocean. When the herring were swimming close to the water's surface, they would throw up a flash of light against the clouds which could be seen by people from land. This belief came from the great importance which in the old days was attached to fishing in the Nordic countries. Our ancestors earnestly followed any omen from the Creator which could alert them to schools of fish heading towards the coast, and in the northern light they hoped that such an omen existed.

In the southwestern part of Finland the northern light was associated with considerably larger creatures in the sea. One belief held that the lights were caused by big whales thrashing waves in the ocean, which in turn created reflections in the sky. In Valdres, Norway, an old tradition related the northern lights to icebergs in the North Sea. Here it was

thud; then stood silent beside the caller, waiting to learn the reason for their summons.

Among Indian tribes in North America it was commonly believed that the northern lights were caused by a gathering of medicine men and warriors in the northernmost part of the world, giving large feasts and preparing their fallen enemies in huge kettles. Other Indians believed that a tribe of tiny Indians inhabited the northernmost part of the ice, several days journey away. They also believed that these small Indians were so strong that they could catch whalefish with their hands, and when they made a fire to cook their catch, the flames of the bonfire were reflected from the sky, thus creating the northern lights.

A related belief can be found among the Maoris, who claim that when their ancestors came southward to the Polynesian Isles some 20 to 30 generations ago, some of them continued on their voyage to a land in the far south where they settled. The Maoris thus believe that the southern lights (aurora australis) are reflections in the sky of huge fires kindled by descendants of those ancient voyagers, who are signalling to their distant kinsmen in the Polynesian Isles, in hope of being rescued from their chilly abode.

claimed that when the northern lights filled the sky, the icebergs were rocking in the North Sea, and therefore windy weather was to be expected from the west.

One of the most romantic conceptions of the northern light can be found in Danish folklore. The tradition was that the northern lights were due to a throng of swans flying so far to the north that they were caught in the ice. Each time they flapped their wings in attempting to free themselves, they created reflections in the sky which could be seen in Denmark as the northern light. A superstition originating in Sweden relates the northern light to the greylag goose, the lights being the greylags flapping their wings.

In Finland the old people named the aurora after an animal and called it revontulet (i.e., fox fires). The tradition behind this name in Finland is that the northern light is caused by a fox with glittering fur running across the mountain area in Lappland.

In this regard it is interesting to note that the oldest known illustration (dating back to 1767) of the northern light from the land of the Lapps depicts a fox-hunting scene. The fox is thus the source of the northern light, while at the same time the northern light provides enough illumination for the hunter to see the fox (see Fig. 1.1).

1.2 The Northern Light – A Vengeful Being

In ancient times most people were afraid of the northern light. For instance, among the Alaskan Eskimos it was common practice to bring the children inside when the mystifying flames of the northern light spread across the heavens. If the children were brought inside too late or not at all, the light would swoop down and grab their heads for use in ball games. If it was impossible to bring the children into the house in time, one could either wave a sharp knife or sing to keep the northern lights at a safe distance. A similar belief can be found in the eastern part of Norway where it was said that the rays of the northern light would descend and chop off the heads of those who kept watching it too long.

The dual theme of northern light and ball games was also well known among the Greenlanders on the west coast, but they believed that dead ghosts were playing ball with walrus teeth. A somewhat more macabre belief is found among the Eskimos in the easternmost part of Greenland, where the rustling flames were attributed to the spirits playing ball with stillborn children or children born in concealment. On the east coast of Greenland the northern light is called "alugsukat", which can be translated as "children born too early" or "in concealment".

Inhabitants of the Faroe Islands considered the northern light a threat to their childrens' life and health, and these were warned not to leave their houses without wearing a cap, for fear that the light would come down and scorch off their hair. This belief was also held in Sweden, where people were warned against having a haircut during displays of the northern light.

One way to prevent the northern light from coming directly to the ground was to stretch some tools or iron towards it.

The Lapps in Sweden warned their womenfolk against leaving the house bareheaded or leaving it at all while the northern light shone. In Norway Lapps

3

also feared that the northern lights would descend to the ground and either steal or kill their children. A well-known tale from Finnmark which is related to the northern light is as follows:

Two brothers went out together to separate their reindeer herd. On this particular evening after they had eaten, the northern light appeared in the sky, and the youngest boy started to mock the light with the following chant:

> The northern light is running lip, lip, lip
> with fat in its mouth lip, lip, lip
> with a hammer in its skull lip, lip, lip
> with an axe on its back lip, lip, lip.

(The word "lip" might be an abbreviation of the word "lipuhit" which means "flicker".) The older brother tried to warn the younger one not to go on mocking, but this only provoked him to become even wilder. The lights then began to agitate frighteningly and a sound was heard as of snow being slapped with a stiff hide. Finally, the northern light came down, killed the boy and burned up his pesk (reindeer jacket); the older boy saved himself by hiding underneath their overturned pulk (reindeer sleigh).

In the Kautokeino area of Finnmark the following rigmarole was used when someone spoke disrespectfully to the northern light:

> The northern light, the northern light
> flickering, flickering,
> hammer in its leg
> birch bark in its hand.

The hammer here is related to the conception of a terrible destructive creature, and the birch bark to its devouring flames.

The more respectful Lapps, however, on becoming aware of an intensely flaming northern light slowed down their speed and removed the bells from their reindeer.

An old tale found in the area of Troms in Norway claimed that anyone laughing at the northern light would be paralyzed since "one was then laughing at the Almighty's power".

In many areas in the Nordic countries it was a widely held belief that the northern light was a vengeful object which killed those who mocked it. One way of teasing it was to wave at it with a white piece of cloth, and by so doing, it was believed that the northern light would come closer, even right down to the ground, where it could cause ruination and disaster. Specifically, in the eastern part of Finnmark it was thought that if a child teased the northern light it would move quickly and finally come down and kill the child.

A well-known tradition both in north and south Norway was the fear of whistling at the northern light. If one did this in Bardu, he would be paralyzed by the light, whereas in Gudbrandsdalen the light would behave crazily.

The fear of whistling at this lively natural phenomenon undoubtedly must be related to traditions in other cultures in the Arctic region, especially those in which it was believed possible to contact dead friends by whistling to the northern light.

In contrast to the situation among the Eskimos where, even up to now, the northern light has been used as an intermediary between life and death; the mythological background has almost been lost in Scandinavia. There remains only fear, maybe only the fear of making fun of the dead.

These few examples probably are rooted in the belief that one should never tease the Almighty; anybody doing that would lose his own power.

1.3 The Northern Light – Flames from the Realm of the Dead

Like the Scotsmen in the old days who imagined the "Fir Chlis" – or the northern light – as an eternal combat between warriors, many other groups of people also associated the phenomenon with violent struggles between heavenly beings.

The Tlingit Indians believed that the spirits of warriors who died in battle dwelled in the sky. The northern light, which was a result of an internal fight among these warriors, foretold catastrophes and bloodshed on Earth. A similar idea was predominant among the Chuchues, who believed that the northern light was the home of those who suffered a violent death.

A story among the Norwegian Lapps says that when two Lapps started to quarrel and the argument became more heated, they sat down on the ground and sang so that the savio – or realm of their benefactors – would give them light, which probably was the northern light. When these two lights met they began to fight. A terrible noise could be heard and a rustling in the sky accompanied these duels among the Lapps. If the light as seen by one of the disputing Lapps slowly faded away he would be ill, and if the light disappeared completely, his sjaman – or benefactor – would die.

Fig. 1.3. A fanciful drawing of a northern light seen in Middle Europe on the 10th of February in 1681 A.D. The sun is below the horizon, and the northern light is drawn as flaming castles and a marching army. The river in the foreground is the Danube

Estonians interpreted the northern light in terms of a fight between heavenly beings. On the isle of Øsel in the Baltic Sea it was related that on some nights heaven opened up and two armed warriors appeared, each eager for a fight, but God would not allow them to fight and separated them.

The Russian Lapps and the Skolte-Lapps also believed the northern light was associated with violence. Among these people the light represented the souls of those who had either been stabbed to death or killed in war. According to the tradition these dead souls lived in a large hall where they quarrelled from time to time and became involved in a big fight in which everyone was stabbed to death, leaving the floor covered with blood; the northern light was a warning that their fight was still continuing. A name for this phenomenon among the Skolte-Lapps was "runtis-jammij" which can be translated as "some who are killed by the use of iron".

It was a fairly common belief among primitive people that the souls of those suffering a violent death were condemned to a restless existence. When one imagines the prominent role played by the northern light on occasions in the sky it is not difficult to understand how its rapid motion and reddish hue could be associated with those who had met a violent death. In some dialects in the Nordic countries the northern light was called "blodlyse" or "blood light" to recall this superstition.

1.4 The Northern Light in Mythology

The mythological role of the northern light was very important in the religion of the Ottawa Indians in Canada, who believed that it was a message from their Creator and benefactor.

According to their tradition the halfgod Nanahboozko created the world and the people. After the creation was completed and man had grown accustomed to his surroundings, Nanahboozko moved to his permanent home farther north. Before leaving his people, however, he promised that he would always look after them and see how they were faring. As evidence of this, he would from time to time ignite mighty flames, whose reflections in the sky would be visible to them. To the Ottawa Indians the northern

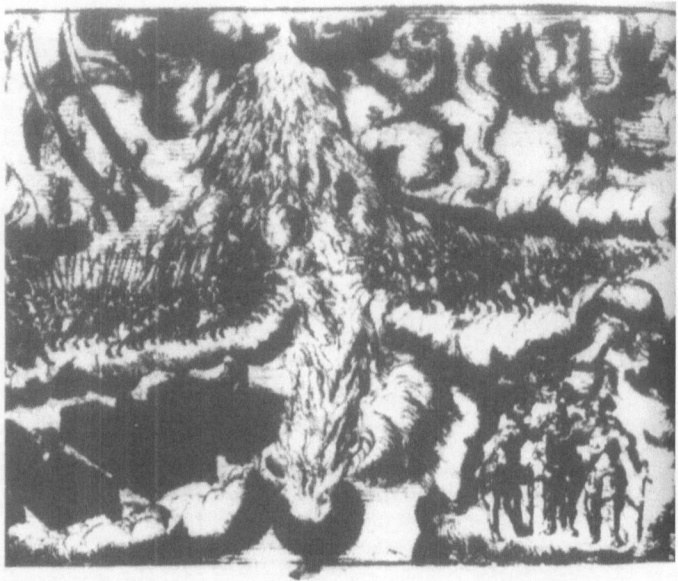

Fig. 1.4. A fantastic illustration of a phenomenon observed on the 25th of November in 1663 A.D. in Hungary. It shows two armies locked in battle, and the northern lights are visualized as a bloodstream from the battlefield. This illustration and the following one is taken from the book *Nordlicht-Beobachtungen in Ungarn* by A. Rothly and Z. Berkes (Akademiai Kiado, Budapest 1963)

light was Nanahboozko's flames, reminding them that their creator and benefactor still cared for them.

The northern light was also very important in the mythology of the Siberian Cuvash tribe. In these peoples' minds there existed a god whom they called Suratan-Tura – which can be translated as "the sky gives birth". The same name could also be used for the northern light, and the tribe thought that when the northern light appeared in the sky Suratan-Tura would give birth to a son. Women giving birth under the flames of the northern lights would be spared the worst pains.

The reddish northern light was contained in the mythology of a group of Russian Lapps. They maintained that a mythological female named Naainas perished when the sun's rays accidentally struck her. The northern lights are created from the blood of Naainas, and her widover is later married to the Sun.

In Finland the northern lights were regarded partly as an admonition against sin, for it was thought that the archangel Michael was fighting the evil spirit Beelzebub by kindling the northern lights. The northern lights were therefore a reminder to fight against one's own sins.

From Vesterbotten in Sweden there is a story which relates the northern light to a similar fight between good and evil. Here the northern light was believed to be due to a fight between Tor and the jotuns. If the light was seen in the southern sky they believed that it was the breathing of Tor and his two rams, whereas if seen in the north it was the breathing of the jotuns in the sky. Tor, who was the Norse god for thunder and lightning, was believed to travel in the sky in a wagon pulled by his two rams, swinging his hammer in the air to produce lightning strokes. The hammer mentioned previously in the Lappish rhyme (Chap. 1.3) may well be connected with Tor.

In Sweden another old tale connects the northern light with a religious superstition since it is said that "God is angry" when the northern light flames.

The Indians of northern Canada also had a deep religious reverence for the northern light. When it appeared in all its splendour in the sky they turned their heads toward it and chanted "Ithenhiela welcomes us to his home behind heaven". With this auroral occurrence they believed that one among them would die and go to the glittering world of Ithenhiela.

Ithenhiela in this legend was a young boy who was taken prisoner by Naba-Cha, the most powerful and cruel human who ever lived. His best friend was a two-year-old moose which rescued him from Naba-Cha's captivity, and Ithenhiela brought from captivity a magic belt which he gave to the great god Hatempka. The belt had marvellous healing powers which made everyone happy again in "the land of heaven". Among these Indians, Ithenhiela was a symbol of godliness, and each time the northern light flamed across the sky, it was thought that it was their god of good will waving a greeting of good wishes to them.

1.5 The Northern Light as an Omen of War, Disaster and Plagues

It was fairly common during mediaeval times to interpret the northern light as an omen of war, conflagration, disasters and plagues. Narratives from all over Europe are embellished with evidence supporting this thesis. At Lista in Norway for example, the northern light was said to predict a war when it came south of the point in the sky where the sun was located at its southernmost position during the day. An old Lista proverb runs: "If the northern light is red, then it is an omen of a coming war". Many other

Fig. 1.5. The description of a phenomenon observed from Bohemia in 1570 reads as follows: "On the 12th of January in 1570 in Bohemia an unusual omen was seen in the sky between the clouds. It lasted for four hours. Firstly a black cloud, almost like a great mountain splattered with many stars appeared in the sky. Above this cloud was shown a strong band burning with sulfur in the form of a ship. From this arose many burning torches, almost like candles, and among these stood big pillars, one towards the east and one towards the north. The flames were running down along the pillars like drops of blood, and the town was illuminated as if on fire. The nightwatch rang his alarm to awake the people so they could see this miraculous omen from God. Everybody was appalled at the sight and said they had never seen or heard about such a terrible sight in their living memory." The text concludes in the following manner: "Therefore, dear Christians, take this awful sign into your hearts and pray that God might ease our punishment"

examples show that the northern light foretold wars; the Tlingit-Indians of southeastern Alaska, for instance, considered the northern lights to be a sure sign of approaching battle.

According to history, the inhabitants of Bergen probably observed a particularly active display of the northern light on New Year's Day in 1702, which was interpreted by many as an ill omen from God. The town of Bergen was in fact reduced to ashes by a conflagration in May of the same year. A similar ominous northern light had probably been observed about three months before the occurrence of another huge, destructive fire in Bergen in 1582.

It is typical that northern lights observed at low latitudes during their rare appearances are predominantly reddish in colour. This frequently led to the belief in southern Europe that large areas in the north were on fire when this frightening phenomenon was observed. Such a northern light was observed from Rome in the year 37 A.D., when the Emperor Tiberius, believing that the village of Ostia was on fire, ordered a troop of soldiers to go and rescue the people who were in danger. Another example of such a "fire" was observed, on September 15th in 1839. A crimson light was perceived in the sky north of London. It rose from the horizon and gradually extended to the centre of the heavens, and by 10 o'clock the whole sky from east to west was one vast sheet of light. It had a most alarming appearance and was exactly like that occasioned by an enor-

mous fire. Thousands of people ran in the direction of the supposedly awful catastrophe. The engines belonging to the fire-brigade stations in Baker Street, Farringdon Street, Watling Street, Waterloo Road, as well as those belonging to the West England station – in fact almost every fire engine in London – were horsed and dispatched to the supposed scene of destruction. Such misinterpretation of the northern light has even occurred in our century.

When Gustav Adolf of Sweden was preparing for war against Poland in 1629, the inhabitants of Danzig saw a sea battle in the air. In the same year in Meissen, Germany, two armies were seen fighting in the sky. They were so easy to separate from the clouds that it was possible to distinguish their features and uniforms, and the largest army lost the battle. In 1563 in the air above Calais, France, two

Fig. 1.6. We find many descriptions of armies fighting great battles in the air which have been related to the northern light. This one refers to a sighting on the 14th February in 937 A.D. The illustration, however, is made by d'Orderius Vitatis in more modern times. Illumination of the ground is noticeable

7

armed warriors, one carrying a cross and the other a lion on his shield were seen. Immediately behind the warriors were two women in long white gowns. People thought that they symbolized the peace treaty which England and France had concluded.

Several descriptions can be found from Middle Europe and the following is an example from 1571 (cfr. Figs. 1.3–1.6).

"Flaming columns were observed above the tower Domztice, and a flaming dragon flew about in the sky. I have heard from my grandfather that this light phenomenon foretells murder, large conflagrations and other horrid disasters".

It appears that many Christians have sought consolation in the Bible and especially in the following words from the prophet Joel whenever they perceived an awesome northern light:

"I will show wonders in the heaven above and in the earth beneath; blood and fire, and pillars of smoke: the sun shall be turned into darkness, and the moon into blood before that great and notable day of the Lord come".

When Martin Luther (1483–1546) witnessed a northern light in 1525 he is believed to have quoted Paul in the following manner:

"Clothe yourself in God's complete armament, so you can withstand the devil's sly attack; because we are not facing a fight against blood and flesh, but against evil forces, against authorities, against the world's rulers in this darkness, against the spiteful spirits in the heavens".

Fig. 1.7. This figure was used by Sophus Tromholt in a paper from 1885 as an example of a strange illustration of a northern light. The drawing was originally published in *The Midnight Cry: Behold the Bridegroom Cometh* by E. M. Hardie (S.W. Partridge & Co London, 1883). The figure illustrates a newspaper cut from The Cincinnatti Sun reading as follows: "On Tuesday night, March 21, 1843, about 11 or 12 P.M., of a sudden light burst forth, the whole face of the earth appearing to be lit up, the light being so vivid as almost to blind the spectators. My first impression was that it lightened very sharply, but as it continued I saw that it must be something else. The captain of the Penn was sitting in the cabin at the time with three or four candles, but seeing the light notwithstanding, and anxious to know the cause, he ran out to the guard and asked Mr. Frances if he saw that light. ‚Yes.' ‚What is it?' ‚Dear only knows, for I don't'. Looking diligently to discover whence this strange light came, they saw in a south-west course, but nearby overhead, the outlines of A SERPENT IN THE SKY, in a crooked position, except the tail, which was straight, and the head towards the east. It then turned to a lively bright red, deep, and awful, and so remained stationary among the stars for several minutes. Then a part of it disappeared to about the middle, and the remainder in a gradual manner formed itself into a distinct Roman ‚G'. In about a minute and a half it changed into an equally distinct "O", in which position it remained about two minutes, when the figure in the heavens again changed to a plain distinct ‚D'. When the O turned into D, it formed a kind of oblong shape, then came straight on one side as a D should be, and having continued in its perfect shape for some minutes it again assumed the oblong shape and disappeared, and the sky gradually returned to its original appearance – Bible Reader".
The event was also mentioned in other Cincinnatti newspapers and evidently the descriptions deal with the occurrence of a meteor rather than a northern light

Many attempts have been made in searching through old annals from all over Europe to find information on the northern light. This kind of research has increased in popularity lately due to an interest in the behaviour of the solar cycle and the relationship between the Sun and weather in historic times. In the following are listed a few such notes from contemporary European annals.

In the year 1104 many people saw a strange sign in the northern sky. The heavens were often set afire, and many stars fell to the ground. Flying tongues of flame, glowing spears and burning torches in the sky were seen.

In midwinter 999 A.D. the heavens opened up and a torchlike phenomenon with a long beam was seen. By the light of this beam, one could not only see objects on the ground, but also inside the houses. When the opening disappeared, the light took the shape of a dragon with a big head and sky-blue legs.

The appearance of a northern light probably occurred at the time when Olav Tryggvason was king of Norway. According to the history written by the Norse author Odd Munch, many of the King's men often saw lights in the sky. Olav Tryggvason was killed in the battle of Svolder in the autumn of 1000 A.D. and shortly before the battle the king's near friend and attendant, Torkjell Dydril, saw a powerful light one night when he followed King Olav ashore, and in the light were two men in beautiful clothing who put their hands on the king's head. It is difficult to date this sight exactly, but it might well be the one mentioned above in 999 A.D.

Of course there are many possible explanations for such an experience described, as this one, by a single person; it might have been the description of an epileptic attack. It is, however, not inconceivable that a northern light could have frightened the men of King Olav Tryggvason.

One can look back even further in history in search of descriptions of light phenomena. In 507 A.D. people in Rome saw glowing armies and bloodshed in the sky, and sounds of trumpets were heard. At this time the Longobards attacked the Roman empire without meeting any resistance.

In 502 B.C. glowing spears were seen one night from Rome. As this occurred shortly before a hostile attack on the Roman empire, where the Roman consul Posthumius lost a great battle, the light phenomenon was seen as an ominous warning of the battle.

One could continue in this way far back into European annals searching for innumerable references to awesome visions in the sky. The illuminations often filled people with terror, and were interpreted as ill omens for war, disasters, plagues, fires and misery. One cannot, without further proof, maintain that all these signs observed in the sky were northern lights, but it is plausible that many of them were.

2 The Northern Light in Norse Literature

2.1 Local Peculiarities are Reflected in Mythology

Assuming that mythology originates in the human aspiration to personify the surrounding natural forces and phenomena, it is logical that natural phenomena, which are of special local importance would be given an important role in the mythology of particular areas.

The contrast between the light summer and the dark winter has left its mark on Norse mythology. The northern light is a phenomenon which dominates our dark winter nights and at the same time creates an impression of mysterious forces and awe-

some features on people who are unaware of its origin. Therefore one would think that this, often violent appearing, phenomenon in our part of the world must have been a dominant element, which stimulated the imagination of our ancestors and led their minds into the realm of deities and mysticism. We could almost take it for granted that this unpredictable light would play a central part in Norse mythology. Curiously enough, the northern light has appeared to have escaped playing a major role in the Norse mythological gallery of personified natural forces and phenomena.

2.2 Norse Literature

In Norse literature written before 1300 A.D. there are two outstanding publications of interest to those seeking knowledge about Norse tradition and belief, namely, *The Poetic Edda* also called *The Older Edda* and *The King's Mirror*. In addition Snorre Sturlasson (1178/79–1241) wrote a book called *The Younger Edda* or *Snorre's Edda* which is a textbook for scalds in Norse mythology. The scalds were the authors, often employed by the Norse kings to compose heroic poems. *Snorre's Edda* is invaluable in interpreting the *Poetic Edda*, but it should be noted that since Snorre lived several hundred years after the Edda poems were written, his interpretations may well be questionable. *The Poetic Edda* consists of 10 celestial and 20 hero-idolizing poems – by unidentified authors. The celestial poems recount ancient myths which are concerned with the accomplishments of the various gods, edited according to their significance in the order of Odin, Tor, Frøy etc; the heroic poems deal with the tragic fate of a few major indi-

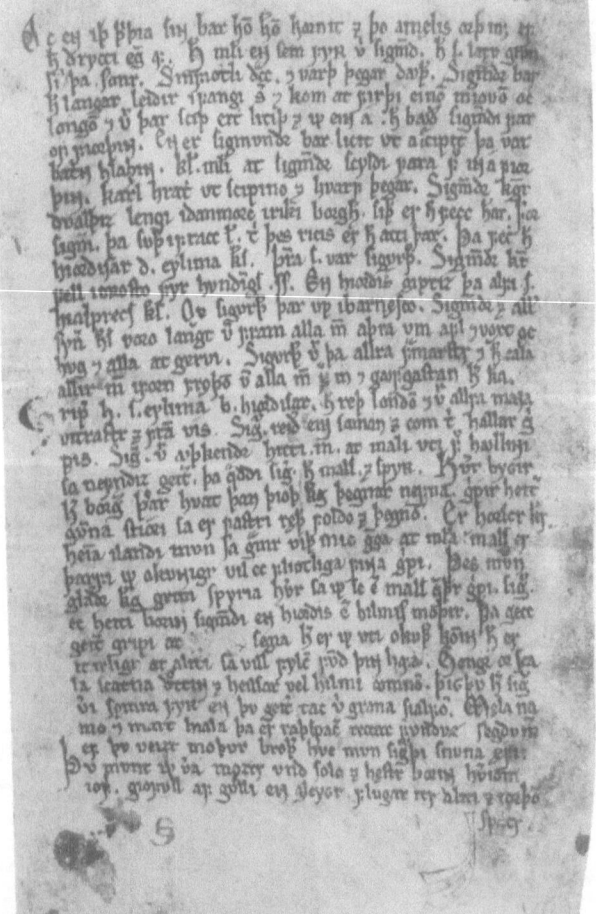

Fig. 2.1. This is a copy of a page from the original handwritten manuscript of the *Poetic Edda* (Codex Regius). Several authors have discussed the possibility of description of the northern light in the Poetic Edda. (Courtesy, The Arnamagneanske Institute, Copenhagen)

viduals. Here much wisdom and worldly experience is displayed. The poems stayed alive more or less intact by oral tradition until they were written down about 1270. They were well preserved in Iceland after the emigration from Norway which began around 700 A.D. Some of them were probably also composed there.

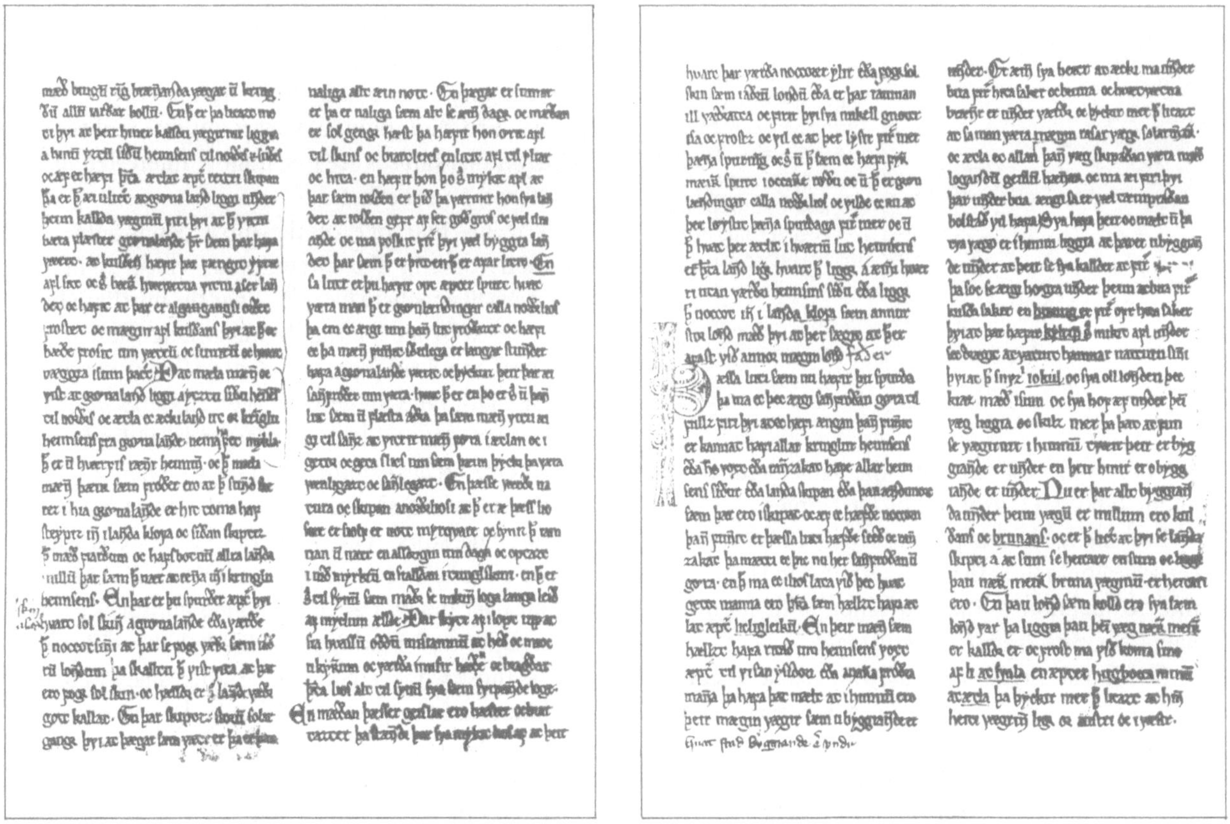

An exact dating of the origin of *Edda* is extremely difficult. Most certainly none of the poems were composed before 700 A.D. and modern research in the field indicates that they were most likely compiled as late as 1000–1100 A.D.

The term Edda is synonymous with greatgrandmother in the old Norwegian language, which probably indicates that the poems were put into writing sometime in the distant past. Form and content are closely interconnected in the *Edda* epics, and it is practically impossible to elucidate the author's meaning exactly in a translation into foreign languages; even translation into modern Norwegian entails great difficulties.

The King's Mirror was probably written before 1250 in the form of a conversation between a father and his son, probably a king and his successor. The son is being taught by his father about all possible

Fig. 2.2. These copies of the handwritten chronicle *King's Mirror* – from about 1250 A.D. – reproduce the description of the northern light which starts in the left column from the 6th line. The text is in Old Norse. (Courtesy, University Library, Oslo)

matters that he would need to know as a future ruler of Norway in the old days. During the conversation the father says:

"But as to that matter you have often inquired about, what these lights can be which the Greenlanders call the northern lights, I have no clear knowledge. I have often met men who have spent a long time in Greenland, but they do not seem to know definitely what these lights are".

Thus, it appears that the author of *The King's Mirror* probably never saw the northern light and that he did not know whether it could occur in Norway at all. On the other hand, he stated that it was known to often occur in Greenland (see also Chap. 4.3).

These few words, more than 700 years old, have been a source of wonder and mystery ever since. In the scholarly literature several authors have neglected this particular point and in fact have suggested that the northern light is a phenomenon which

Fig. 2.3. Anne Holtsmark (1896–1974) professor in Norse mythology at the University of Oslo from 1949. She has published several articles and books about Norse mythology and also translated into Norwegian some of the original Norse material

did occur in Norway with the same frequency at about 1200 A.D. as it did at later times.

Leaving aside this peculiarity in *The King's Mirror*, we should note that the author is in fact the very first to mention the phenomenon with its original name *Nordurljos* or *Northern Light*.

In Finnish mythology, as with religious conceptions in so many other societies, a river constitutes the border between the realms of the living and dead. The name of the river was Rutja, which stood in fire and was related to the northern light. In Norse mythology the corresponding river is called Gjoll, the reverberating river, according to Snorre's *Edda*, and crossing this river was the bridge called Gjallarbroen guarded by Modgunn. This bridge is covered with glittering gold and will give way on Doomsday. It is tempting to draw a parallel between the Finnish river Rutja and the Norse river Gjoll, and based on this similarity, Gjoll can be seen as a metaphor for the northern light. A trembling auroral arc can also well be associated with a glittering bridge.

Allusions are found, especially in non-Scandinavian literature, that Bifrost, the bridge leading from the Earth to the heavens, is a metaphor for the northern light. This might be a superficial use of original sources, as it is claimed in Snorre's *Edda* that Bifrost is the rainbow. Even so, scholars of Norse mythology today do not quite agree on what the word Bifrost actually means, whether it is "the trembling road" or "the road with three colours". One well-known interpreter of the *Edda* poems in Norway, Anne Holtsmark (1896–1974)[1] was familiar with the idea that Bifrost represented the Milky Way, whereas the northern light did not appear to interest her in this connection. There are, however, surprisingly few scholars of Norse mythology who are interested in the influence which local natural phenomena might have had on the authors of the *Edda* poems. From our point of view it appears as if the experts on this subject are much more interested in semantics than substance.

It seems that one of the most commonly held opinions among the circumpolar cultures in the northern hemisphere has been that the northern light was caused by an external battle among those in the realm of the dead who had met a violent death. In Norse mythology these were the "einherjes", or the fallen warriors and heroes received by Odin in Valhalla. One must therefore wonder whether the "einherjes" played a similar role on the horizon of our Nordic ancestors. The Valkyrjes, who were young maidens, rode on horseback out of Valhalla through the air over the battlefield, bringing messages of war from Odin and marking, on his behalf, with their spears those who were to fall in battle. They were the harbingers of war for the Vikings and the messengers of battle for the "Einherjes" – and therefore played the same role to the Vikings as the northern lights did to the Skolte – Lapps (cf. Chap. 1.3).

The modern experts are in doubt as to whether or not the northern lights played a central role in Norse mythology in spite of frequent allegations by earlier authors that the northern light was interpreted by the Norse people as reflections from the shields of the Valkyrjes.

There are, as already stated, two main sources in which the Norse mythology can be studied; the *Older Edda* and the *Younger Edda*. These works have been the subject of many different interpretations, especially as regards the poetic part. It is also interesting to observe how these interpretations have been influenced by the cultural style in vogue at the time they were made.

How the *Edda* is interpreted and understood will often be a question of background knowledge and associations of the interpreter and this in turn will be strongly influenced by the school of thought prevalent at the time of analysis. One must remember also that the more specialized our society becomes, the more difficult it is to correspond across professional barriers. Problems interesting for scientists in one field may be uninteresting and completely overlooked by those in another. To scholars particularly interested in Norse literature, the northern light is of no special importance, and therefore it is possible that it has been overlooked in this regard.

[1] Anne Holtsmark: Studier i Snorres Mytologi, Universitetsforlaget, Oslo 1964

Fig.2.4. This figure illustrates a Valkyrje serving the warriors before bringing them to Valhalla. It has been suggested that the northern light was related to the Valkyrjes in Norse mythology. For more details see the text in Chapter 2

It was especially in stanzas 23 and 24 in *Voluspå* or *The Prophecy of Seeress* that Magnusson found the most convincing evidence for his belief that the Valkyrjes originally must have been metaphors for meteors and northern lights. The stanzas read as follows in Hollander's[2] translation into English:

23 The Valkyries flock
   from afar she beholds,
   ready to ride
   to the realm of men:
   Skuld held her shield,
   Skogul likewise,
   Guth, Hild, Gondul,
   and Geirskogul:
   [for thus are hight
   Herian's maidens,
   ready to ride
   O'er reddened battlefield].

24 I saw for Baldr,
   the blessed god,
   Ygg's dearest son,
   what doom is hidden:
   green and glossy,
   there grew aloft,
   the trees among,
   the mistletoe.

The comments by Finnur Magnusson to these stanzas are as follows:

"Originally the Valkyrjes were certain meteors or phenomena in the air like fireballs, flaming northern lights etc. They were sent from Valhall, in other words the vault of heaven, by Odin, the upper deity. Still the plebeians in many countries believe that such phenomena signifies coming wars and calamities. Here the death of Balder is announced by their departure".

2.3 The Northern Light – Reflections from the Shields of the Valkyrjes

Between 1821 and 1823 the Icelandic author Finnur Magnusson (1781–1847) published a Danish translation with comments on the *Older Edda*. Magnusson was rich in ideas and associations, and he was widely read. Certainly his opinions were strongly coloured by national romanticism which began to dominate contemporary thought at the beginning of the last century. To Magnusson it was important to emphasize nationalistic aspects in his interpretation of *Edda*. He was looking for influences of a special Nordic nature in the poems, and maintained that the northern light was mentioned in *Edda* by such metaphors as reflections from the shields of the Valkyrjes and the Gjallar bridge. He also claimed that all the different characteristic names of horses used by the gods and the Valkyrjes' names that referred to light flash, flames and fires, actually were references to the northern light.

In the last century many authors agreed with Magnusson in his interpretation of *Edda*. It was indeed a very popular understanding that the *Edda* poems were inspired by original Nordic motives, and that they were very little influenced by universal mythological accounts.

2.4 The Saga Literature (Old Norse Prose Narrative) Describes Many Light Phenomena

To the Saga literature belong the *Saga of the Nordic Kings* by Snorre Sturlasson and the *Sagas of the Icelanders*, among others. Most of the Saga literature was written by unknown authors between 1100 and 1400 A.D. The subject of the Saga literature can either be the life of the kings or some outstanding Icelanders living between 850 and 1050 A.D. These Sagas were handed down through the centuries by oral tradition from one generation to the next before finally being put into writing.

---

[2] Lee M. Hollander, The Poetic Edda, University of Dallas Press, Austin, Texas 1969

Fig. 2.5. Gerhard Munthe (1849–1929) made this vignette for a modern issue of the *Saga of the Norwegian Kings* written by Snorre Sturlasson (1179–1241) which shows Vikings sailing underneath arcs of the northern light. Snorre Sturlasson, however, never mentioned this phenomenon in his famous chronicle. (Kongesagaen by Snorre Sturlasson, translated by Anne Holtsmark and Didrik Arup Seip, Gyldendal Norsk Forlag, Oslo, 1975)

In the Saga literature many light phenomena, in particular, are referred to before the occurrence of important battles. One of the best-documented eclipses in Norway during the Viking era occurred during the great battle of Stiklestad in 1030 A.D. when King Olav Haraldsson (St. Olav) was killed. Another light phenomenon is mentioned in *The Njål's Saga* in the poem *Darradarljod*. This poem depicts a vision that Darrad had early in the morning of Good Friday in 1014 A.D. before the battle of Brjån at Cloutarf in Ireland. The 9th stanza reads as follows in our own translation into English:

> Frightning it is
> to look around
> where gloomy blood-
> coloured clouds,
> are drifting across the sky
> The air is red
> dyed by blood
> So well the
> Valkyrjes could sing.

It is very likely that this light phenomenon – where the air is dyed with blood – could be the northern light. In Scotland where Darrad lived, the northern light when seen is often red. The experts in Norse literature generally maintain that this poem was inspired by the light of dawn. It is, however, surprising to notice how little fantasy the philologists display in matters such as these; it seems that the only natural red-coloured phenomenon which they have observed in the sky is the light of dawn.

In the *Edda* poem *Skirnesmål* Frøy had taken a seat on Odin's own throne in Lidskjalv to overlook the universe. Frøy rules over weather and crop, wealth, happiness and peace. From this seat he had the opportunity to see the northern light. Far in the north he saw Gjerd Gymesdatter, a beautiful daughter of the jotun Gyme, the Frost giant, as she walked across her father's farm yard in Jotunheimen. Frøy fell in love with Gjerd, and ordered his valet Skirne to go and propose to her for him. Finnur Magnusson in his comments about this poem maintains that Gjerd here represents the northern light. Frøy is, in his opinion, a metaphor for the Sun and the valet accordingly represented the Sun's rays. It is in the following stanza that Magnusson found his most convincing arguments for such ideas (translated by L. M. Hollander, in *The Poetic Edda*, 1969):

> From on high I behold
> in the halls of Gymir
> a maiden to my mind;
> her arms did gleam,
> their glamour filled
> all the sea and the air.

In other words, when Gjerd walked across the farm yard and lifted her arms to the heavens they were shining, and the light from them covered the sky. It is this light that Magnusson brings into association with the northern light.

The Swedish author Esaias Tegnér (1782–1846) (cf. Chap. 3.2) probably had similar associations in mind when he wrote the following stanza in the poem *Frithiof and Ingeborg*:

> Why are Gerda's cheeks praised so much
> the northern light on new-fallen snow touch?
> I have seen cheeks: One day igniting
> two red lights of dawn together.

The Norwegian author Henrik Wergeland (1808–1845) probably also thought he saw similar associations when he mentioned Freja (Frøy) in this poem *Stormen* (The Storm).

> Northern light ride the
> winters dapple-gray mount.
> – Oh silent it just stood
> underneath the dusk's
> scarlet and shook its mane.
>
> The northern light ascends
> on the evening clouds?
> see each like a beauty,
> even to Freja's satisfaction!

Fig. 2.6. This old bronze figure probably illustrates the Norse fertility god Frøy who was one of the most important gods in Norse mythology, together with Odin and Tor

daylight returns to the Arctic region, and the northern light will be devoured by the Sun, just as Gjerd is blended in with Frøy.

This juxtaposition of metaphors was later strongly attacked, and in particular by Magnus Olsen (1878–1963)[3]. Olsen maintains that in this poem Frøy represents the Sun and Gjerd the power of germination in the seed: the two merging together to produce a crop. According to Olsen, this interpretation is strengthened by the fact that Gjerd and Frøy meet in "Barrelunden" – a name which can be translated as the barley grove.

Surprisingly enough Magnus Olsen pays little attention to the light phenomenon itself. He argues that *Skirnesmål* refers to a fertility ritual, in which the light underneath the arms of Gjerd points towards "kornmo" or flashes of heat lightning. Flashes of heat lightning are due to distant lightning often seen in the autumn in open country as for example, in large fields of barley. For a very long time people have believed that these flashes of heat lightning are related to the ripening process of barley.

There is an old tradition in Medelpad, Sweden, which maintains that "when the northern light is burning, the seed will be abundant". Therefore it is not unusual in some areas of the Nordic countries to associate the northern light with the ripening of the crop. The light described in *Skirnesmål* can therefore represent the northern light in a slightly different way than Magnusson maintained. However, it is surprising that so little attention is paid by modern researchers to that part of *Edda* mythology in which light phenomena such as this are described. They focus their interest in linguistic arguments rather than considering relationships with natural phenomena.

Peter Andreas Munch (1810–1863), a well-known Norwegian historian, first introduced a modern interpretation of the *Older Edda* in his book *Nordens gamle Gude- og Heltesagn* 1841[4]. Munch disassociated himself from Magnusson's tradition of national romanticism. It seemed unthinkable to Munch that the northern light in any way could be related to the Valkyrjes. The northern light is not mentioned at all in Munch's interpretation of *Edda*.

There are other stanzas in *Skirnesmål* which Magnusson maintained could be metaphors for the northern light. In the 8th stanza, after Frøy had asked Skirne to go and propose for him, Skirne answered as follows:

Thy steed then lend me
to lift o'er weird
ring of flickering flame,
the sword also
that swings itself
against the tribe of trolls.

The words "visan vafrloga" in particular are the ones which Magnusson maintained are a metaphor for the northern light. He translated these words into "flickering flames" which is in very good agreement with modern translations of the same stanza. After all, what flames in the night sky flicker more than the northern light!

To Magnusson the love story in *Skirnesmål* is a portrayal of the Sun – represented by the god Frøy – who wishes to light up the dark polar regions where the jotuns dwell, the frost giants, and the evil forces of the darkness. In this area lives the beautiful maiden Gjerd (the northern light) who is the daughter of the jotun Gyme (the Arctic Ocean). Frøy, the Sun, being unable to reach this region with his rays during winter, asks his valet, Skirne (the one who cleans up the air) to go ahead of him. Gjerd is willing to meet Frøy after 9 days. This delay of 9 days, Magnusson claims, refers to the fact that after 9 days,

---

[3] Magnus Olsen, in Gammelnorsk Myte og Kultus, Maal og Minne, 1909
[4] This book has been translated into English. P. A. Munch, Norse Mythology, Oxford University Press 1926

Interpretation of Norse literature is obviously a difficult task. People such as Magnusson, Olsen, and Munch with different backgrounds and abilities may reach vastly different conclusions in associating the *Edda* poems with natural phenomena. Another example of disagreement in translating or interpreting words from the Old Norwegian language is the word "Hålogaland". Hålogaland is the Old Norwegian name for the northern part of Norway. Sophus Bugge (1833–1907), a Norwegian linguist, in 1871[5] claimed that the word Hålogaland is composed of the prefix "hå" which means high in the air or aloft, and the word "loga" meaning flames or blaze. The full translation of the word "Hålogaland" would then read "The land under the flames high in the air" or "the land under the northern light". Later Halfdan Koht (1873–1965), a Norwegian historian and politician, in 1920[6] discussed the meaning of the same word and refuted Bugge's interpretation; he even claimed that Bugge in a lecture had rejected this theory in the 1890's.

2.5 The *Edda* Poems as Referenced in International Literature on the Northern Light

It is amazing that people are so willing to accept the more easily intelligible interpretation of Norse literature as compared to that of an expert on *Edda* mythology. The national romantic school introduced by Magnusson to the studies of the *Edda* literature was fairly popular in Europe in the last century. We find influence of this school in books such as:

Das Polarlicht, which the Swiss scientist Herman Fritz (1830–1893) published in 1881. This book contained one of the first detailed comparisons of the enormous amount of observational material which existed near the end of the last century. In the introductory chapter of his book Fritz quotes some verses from the *Older Edda* which he maintains are descriptions of the northern light.

Ein zehntes (Lied) verwend' ich, wenn durch die Luft
    spukende Reit'rinnen (Nachtmaren, düstere Abbilder
    der Walkyren) sprangen.   (*Rûnatals-thâttr*).
Glut seh' ich leuchten und lodernde Lohe.
                      (*Hyndluliódh.*)
Doch Zeit ist zum Ritt auf geröthetem Wege:
Den Flugstieg lenk' ich das leuchtende Ross;
Muss sein im Westen der Windhelmbrücke (Milchstrasse)
Eh' Walhall's Sänger (der Hahn) das Siegervolk weckt!
          (*Helgakvidha Hundingsbana.*)

The first stanza is verse 155 from *The Sayings of Hár* (Håvamål). *The Sayings of Hár* is a celestial poem consisting of five parts. Stanza 155 is from the last section, wherein Odin relates the 18 magic songs which he possesses. The stanza reads as follows in the English translation by L. M. Hollander[2]:

> That tenth I know,
> if night-hags
> sporting I scan aloft in the sky.

It is presumably the word "night-hags" or witches in the sky which made Fritz think of northern lights in this context. There is, however, nothing in this stanza which indicates that the author has been inspired by northern lights except that the phenomenon occurs in the night sky.

Another phrase that is included in Fritz's book is stanza 49 in *The Sayings of Hyndla* (Hyndleljod):

> A fire see I burn,
> flameth the earth.

This poem is not very well preserved and is therefore difficult to translate into modern languages. Undoubtedly this accounts for the diverse interpretations. Even if the aurora was often associated with terms like "fires" and "flames" during the Middle Ages, no scholar today would dare maintain that this is necessarily a description of the northern light. On the contrary, a more reasonable conclusion is that the poem probably refers to a volcanic eruption. The fact that the poem was probably written in Icelandic also makes this likely since Iceland is a land of volcanoes.

The third quotation included by Fritz constitutes verse 49 in *Second Lay of Helgi the Hunding-Slayer*. This tells of Helgi who killed King Hunding and later fell in battle, thus arriving in Valhalla to become Odin's right-hand man. When morning dawned, Helgi arose and said:

---

[5] S. Bugge, Tillægsbemerkning om navnet Halgaland, Helgeland, Historisk Tidsskrift, Kristiania, 1, 136–140, 1871

[6] H. Koht, Om namne Hålogaland, Haaløygminne, 3–11, 1920

Fig. 2.7. This copy which is taken from the book *Das Polarlicht* by H. Fritz from 1881 shows a German translation of a few stanzas from the *Poetic Edda*. These quotations are according to Fritz and others thought to be descriptions of the northern light. For a more complete discussion of this see the text in Chapter 2

Along reddening roads to ride
I hie me,
on fallow steed aery paths
to fly:
to the west shall I of
Windhelm's bridge,
ere Valhall's warriors
wakes Salgofnir.

It is "the reddening roads" (roðnar brautir) that Fritz here referred to as the northern light. From the end of the stanza it is clear that the story takes place in the early morning hours, since Salgofnir is the rooster of Valhalla which wakes up the warriors. "The reddening roads" are therefore more likely to be a metaphor for the red light of dawn.

The conclusion must therefore be that Fritz's interpretation of a few stanzas in the *Poetic Edda* were not carefully investigated, and cannot be taken as proof of the northern light being described in one of the oldest historical documents of Norway.

2.6 Impact of Fritz's References to the *Poetic Edda*

From the evidence above, it appears that Fritz was not a careful interpreter of the *Edda* poems. For artistic reasons, poets often make use of imaginative metaphors and therefore poetry must be read carefully and objectively when used as a source for relations to scientific matters. This is even more true when poems are translated. Fritz, in a vignette to his introductory chapter, quotes a German translation of a verse in the 14th song of *Frithiof's Saga* by Esaias Tegnér (cf. Chap. 3.2). The verse reads as follows in the German translation:

Du, Nordlichtkrone,
Du, hellst die Nacht
Der nord'schen Zone
Mit Rosenpracht;
Umströmst mit Flüssen
Von Gold den Pol!

In the Swedish original of *Frithiof's Saga* this verse does not exist. Evidently the German translator was of the opinion that this magnificent poetic work should have included some lines about the northern light. It appears as though Henry W. Longfellow (1807–1882) must have used the same translation, since he in fact quotes the same verse in English (cf. Eather 1980) [7]. Longfellow was inspired by German

romanticism and most likely used the German translation instead of the Swedish original. In his poems, Longfellow also asscociated the northern lights with the god Tor. This interpretation by Longfellow may be due to his romantic fantasy although there is some evidence for this connection in the old tradition from Vesterbotten in Sweden (Chap. 1.4) and also in the Lappish rigmarole (Chap. 1.2).

Another point that strengthens the impression that Fritz was not too careful about his use of foreign literature as sources for his work is his interpretation of some lines in *Germania* by Tacitus (55–117 A.D.). The lines taken from Chapter 45 about the Arctic Ocean read as follows:

"Trans Suionas alind mare, pigrum ac prope immotum quo cingi cludique terrarum orbena hinc fides, quod extremus cadentis iam solis fulgor in ortus edurat adeo clarus, ut sidera hebetat; sonum insuper emergentis audiri formasque equorum et radios capitis aspici persuasio adicit".

This can be translated into English as follows:

"Beyond the land of the Svioner is another sea, calm even to stagnation. That the circle of the earth is surrounded by this sea is likely for the reason that the last gleam of the setting sun lingers till he rises again with such brightness that it dims the stars. People with much knowledge maintain that one in fact can hear the sun-god when he emerges from the depth and make out the contour of his horses and the rays around his head".

Fritz maintained that the light phenomenon described here as lasting from sunset till sunrise refers to the northern light. This is rather questionable as it is more likely that Tacitus is thinking of the midnight sun. Furthermore Fritz claims that the sound mentioned by Tacitus is that which some people believe they hear when they see a northern light. Ancient people also believed that when the Sun was setting in the ocean a hissing noise could be heard as when a hot iron is put into water. It should be noted that this is the only place in ancient literature where any sound is mentioned in connection with sunrise or sunset. The English word day-break like the German Tagesbruch may well refer to the same meanings.

In order to be fair to Herman Fritz we must admit that when he claimed that the Norwegians believed they saw the Valkyrjes and heard their song in the northern light, he was referring to the book *Die Religion der alten Deutschen* by Georg Christian Braun, which was published in 1819. This book, written for more advanced scholars, contains the following lines about the relationship between the northern lights and the Valkyrjes:

„Die Norweger glaubten im Glanze des Nordlichts, das manche Gestalten zeigt, die Walkyrien zu erblicken, und deren Ge-

---

[7] Robert H Eather, Majestic Lights, American Geophysical Union, Washington D.C., 1980

sänge im Geräusch der elektrischen Luft zu vernehmen. Dies gibt auch einen Wink zur Erklärung der Stelle des Tacitus (Germania 45.) vom Klirren des Sonnenwagens. Und wenn man statt *formas deorum* liesst: *equorum*, so kommt die alte Idee von den Walkyrien heraus, die in der Luft reitend, und von Flammen umgeben gebracht wurden."

The ideas that Braun presents here are the same ones which Magnusson discussed a few years later and were certainly inspired by Romantic School. Students of Norse mythology today reject Braun's and Magnusson's ideas entirely. One must wonder, however, whether these two authors, living almost two centuries ago, actually experienced this same reaction from their contemporary Norwegians. In 200 years, traces of such a tradition could well vanish, but it may explain Fritz's statements that the Norwegians saw the Valkyrjes and heard their songs in the northern light. At the time these two authors published their books, the activity of the northern lights was probably low (Fig. 2.8). It is likely that in periods of time when the northern lights appear infrequently, such superstitions actually could grow when the northern light *did* appear. In any case it does not prove that the Vikings who lived a millenium earlier actually had the same belief.

As demonstrated above, it appears that Fritz did not make a very thorough analysis of the *Poetic Edda* when he quoted a few verses as evidence for metaphors of the northern lights. In spite of this his conjectures have inspired scientists of our generation. In W. Petrie's book *The Story of the Aurora Borealis* [8] (1963) it reads the following:

> "One can produce arguments that suggest Bifrost was the aurora. The word means 'vibrating way' or 'trembling way', indicating that the bridge was in motion. Furthermore, Bifrost was sometimes described as a green object, and this is the colour characteristic of an auroral arc".

To illustrate this, Petrie quotes from a poem by the Danish poet Adam Oehlenschläger (1779–1850):

> Bifrost i' th' east shone forth in brightest green,
> On its top, in snow-white sheen,
> Heimdal at his post was seen.

Neither in the *Poetic Edda* nor in the *Snorre's Edda* is the green colour of Bifrost mentioned. Oehlenschläger's description is most likely a result of the artist's imagination.

Petrie's book and *Das Polarlicht* are the ones most often quoted with regard to the history of the north-

ern light, but the few lines referred to above indicate that Petrie had not made a very thorough study of modern *Edda* research.

Nevertheless, Petrie brings up a point which deserves a closer examination. As already mentioned above (Chap. 2.2) Anne Holtsmark was in favour of another interpretation of this word Bifrost, namely a road with three colours, and she did not relate it to the northern light at all. In this regard Holtsmark follows many other recent expert interpreters of the Norse literature and mythology who lack the imagination and inspiration to look for natural occurring celestial phenomena in the enormous gallery of persons, creatures and objects by which the Norse mythology is so enriched.

One obvious reason for this complete lack of physical interest among the interpreters of Norse literature and mythology is that only scholars with a background in religion and language appear to have taken a profound interest in this very valuable historical material.

In Snorre's Edda, Snorre himself says that Bifrost represents the rainbow. Because of Snorre's great authority, hardly any one has dared to dispute his interpretation, in spite of the fact that Snorre also relates Bifrost to a burning fire above the bridge and that a very plausible interpretation of the word Bifrost is a "trembling road". Based on these facts that Bifrost is a natural light phenomenon in the sky simultaneously burning and vibrating, it becomes much more likely to associate it with the northern light than the rainbow. This interpretation has been suggested before but never accepted by the expert in Norse literature. There may be several other words in the Norse vocabulary by which a more thorough analysis can be related to the northern light. Hopefully someone with a more multidisciplinary background can take an active interest in this field in the future.

2.7 Traces of the Northern Light in *Edda* Literature

In his book *Majestic Lights* Eather makes the following statement: "The rich Norse mythology associates the northern light with the most beautiful of all goddesses, Freya, goddess of beauty and love". It is not Freya but her brother Frøy (the fertility god) who is described in the poem *Skirnesmål* as falling in love with Gjerd, the beautiful daughter of a giant. As has

---

[8] William Petrie, Keoeit, The Story of the Aurora Borealis, Pergamon Press, 1963

already been mentioned, Magnusson maintained that Gjerd represented the northern light while Frøy was a personification of the Sun.

If one takes time to study more closely the *Poetic Edda* in search of a stanza which can be an allegory of the northern light, he finds very little fruit for his effort. We come to the conclusion that the northern light is never mentioned by name; if it is described, it must be in some form of euphemism. On the basis of modern *Edda* research it may be claimed that some verses are metaphors of the northern light. Two of the stanzas which probably inspired Braun to relate the northern light to the Valkyrjes are stanzas 15 and 16 in the *First Lay of Helgi the Hunding-Slayer* (in Hollander's translation):

15 A light shone then
   from Loga Fell,
   and out of that light
   lightnings flashed:
   (saw the matchless hero
   the maidens riding)
   high and helmeted
   to heavenly realms.

16 Were their byrnies
   blood-bespattered,
   from thin spear points
   bright sparks flew forth.

These are examples of the stanzas in the *Poetic Edda* which describe the light rays from the Valkyrjes' spears. The maidens ride down against a background of gleams and lightning emerging from a nearby mountain. As this light phenomenon is related to the mountain, it is more natural to associate it with the eruption of a volcano than with a northern light.

In stanza 23 in the *Second Lay of Helgi the Hunding-Slayer* an event is depicted which may have been a northern light (in Hollander's translation):

23 What king is it
   these keels who steereth?
   His golden banner
   at the bow floateth,
   his proud prows seem
   no peace to betoken
   a blood-red glow [9]
   forbodeth war.

It is the old Norwegian word "vigroða" which Hollander has translated to "a blood-red glow". Literally this word can be translated to "warfire", which was an omen of war and trouble, a common belief also related to large displays of northern lights. The original text at this point reads "verpr vigroða", where the meaning of the word "verpr" is "to

throw", and no flames in the sky appear more as if they have been thrown there than the northern lights. A common translation of this phrase "verpr vigroða" into the more modern language is "lekende vêrljos" (playing weather-lights) and "vêrljos" is a well-known word for the northern lights (cf. Chap. 8). It is, however, probably more natural to associate the word "vigroða" with the light of dawn, as Hollander's translation also seems to indicate.

Another stanza which can be associated with the northern light is found in the *Baldr's Dream* (Vegtamskvadet), again in Hollander's translation:

Cease not, seeress,
till said thou hast:
answer the asker
till all he knows:
who are the girls
that greet so sore,
and their kerchiefs' corners
cast to the sky?

Here Odin asks Volva (a woman thought to be able to confer with the gods through magic and to foretell the future) and gets no answer, since Volva is not able to tell. This stanza has been interpreted in several different ways and it might as well be associated with the northern light and the old belief in Norway that the northern light was old maids "dancing in the sky", or "dancing maidens waving their mittens" (cf. Chap. 1.1). Where it is said that the maidens throw their kerchiefs towards the sky, it may simply be a metaphor representing a dance, and since "dancing maidens" have been used as a notation for northern light it is not unlikely that these have been the associations of the composer. The most probable translation of this phrase which reads in the original language "ok á himin verpa hálsa skautum", is the one that Hollander mentions in a footnote:

"Probably, there is a pun intended, for the words of the original here translated 'kerchief corners' may also mean 'the corners of the sail'. To account for the riddle being introduced here it has been suggested that the lines refer to the sail of the ship bearing dead Baldr's body".

Another part of the poem with similar connotations is "that greet so sore". The original text here reads: "er at muni gráta". This is rather difficult to translate precisely but it associates the maidens with "hot crying". The Norwegian forefathers of many years ago were not able to distinguish between differ-

---

[9] Hollander makes a footnote here: "Most likely, from the red warshields" (cf. footnote 2)

19

ent light phenomena in the sky and so called them all by the name meteor – meteor might well be poetically described as "hot crying". The composer of this lay might have mingled these different light phenomena, and therefore "hot crying" and "dancing maidens" could both represent the northern light.

We believe, however, with Hollander, that this poem more than any other one probably describes ocean waves. Waves in Norse literature were often depicted as daughters of the jotun Ægir, the giant of the ocean. The cries of the maidens are then represented as droplets of the surf which encircle Baldr's boat.

In *The Sayings of Fåvne* there are two stanzas (42 and 43) describing both a Valkyrje and a light phenomenon. It reads as follows in the translation of Hollander:

A˙high wall standeth	A Valkyrje rests
on Hindar Fell	on the rock in steep,
all enfolded is it	flickering fire
by fire without;	flames about her;
cunning craftsmen	with the seep-thorn Ygg
this castle builded	her erst did prick:
of the glistering	other heroes she felled
gold of rivers.	than he had willed.

The big hall in the Hindar mountain which is surrounded by flames reminds us of the stories about big castles being seen in fire in Middle Europe when instead people probably were seeing the northern light. The flames playing above the Valkyrje as she sleeps in the mountain may well be a metaphor of the northern light, and if true then this is another example showing that the northern light is a warning sign. In this case, it is the Valkyrje's fate which is foretold. Since the light and the Valkyrje are mentioned together, it might mean that they had some relationship, in which the Valkyrje alerted herself about her coming death by the northern light.

2.8 In Norse Literature –
Only Meager References to the Northern Light?

As mentioned above it is very difficult to find any basis for maintaining that the northern light has been mentioned in *Edda* literature. One can only speculate, and most of the different interpretations that have been made in relation to the northern light are often very loosely woven. It is tempting to claim that the northern light probably never adequately inspired the scalds[10] to compose a poem to its glory. It is difficult to overlook the fact that the northern

light is one of the most fascinating phenomena occurring in the polar sky – fascinating for those sensing its beauty, but terrifying to those not perceiving its true nature. The *Edda* poems were composed by people who knew nothing about the cause of the northern light, and as with other inexplicable natural phenomena, people were afraid it was either a sign that their gods were angry or that the devil was looking for revenge.

It has been suggested that our Nordic ancestors did not have an eye for such a beautiful sky spectacle, and that they in fact tried to avoid noticing it. One cannot accept this as an explanation for the lack of references to the northern light in the *Edda* poems.

The reason for this dire lack of references must be sought in other arguments, and the explanation might be closer than we realize.

In this regard it is very interesting to notice how the author of *The King's Mirror* refers to the occurrence of the northern light (cf. Chap. 2.2).

The author of *The King's Mirror*, who lived in the northern part of Norway just south of the Polar Circle at about 1230 A.D., appears never to have seen the northern light himself, and is uncertain whether it could be seen in Norway at all. It is surprising since the author discloses such great knowledge about so many diverse subjects. He also quotes most of the contemporary literature which shows that he was widely read. On the other hand the author knew that the northern light occurred in Greenland.

One possibility is that he personally had not seen the northern light, but was only relating stories he had heard about it from people who had visited Greenland. In that case, there are good reasons to believe that the northern light was an uncommon phenomenon in Norway in the middle of the 13th century. Since the northern light was known in Greenland, however, it must be concluded that the northern light occurred in an auroral zone which was positioned differently than is the current auroral zone.

The other possibility is that the author never saw the northern light, and furthermore, he had probably not met anyone who had seen it from Norway. If he had, his wording would have been different. The reason for this may be because it mainly occurred farther in the north of Norway than where he was ac-

---

[10] Norse name of the composers of the *Poetic Edda* and other poems written in old Norwegian

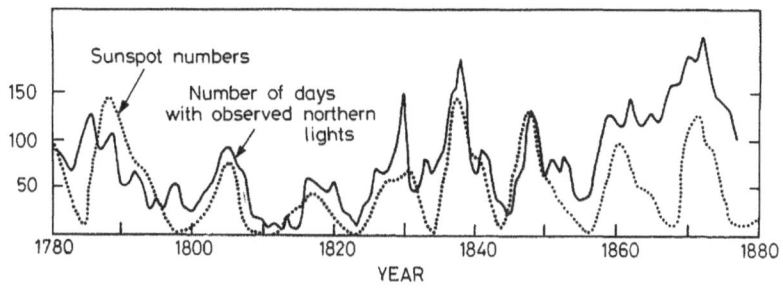

Fig. 2.8. Tis figure shows the number of occurrence of sunspots and northern lights observed from Norway between 1780 and 1880. Note the high correlation of the two phenomena between 1780 and 1850. The figure is based on data collected by Sophus Tromholt about 1880

tually living. The name "nordurljos" relates to this northerly position, and because the author thought it had its origin in Greenland, he talked about it as a Greenlandic phenomenon.

Whichever way one chooses to interpret the wording in *The King's Mirror*, the northern light probably occured more to the north in Scandinavia at that time than it does today.

The occurrence and position in space of the northern light are mainly determined by two factors: the disturbance level in the Sun, and the strength and orientation of the Earth's magnetic field. These phenomena will be discussed more thoroughly in Chap. 10. Here we will only mention that solar activity changes drastically from time to time, and in general one can say that when activity is high, the chances for the northern light to occur are also high. One of the first physicists to discover this relationship was the Danish-Norwegian school teacher Sophus Tromholt (1851–1896), and the diagram shown in Fig. 2.8 is constructed using Tromholt's data. A close correlation is seen between the occurrence of sunspots and the northern light in Norway. It is also apparent from this diagram that both activities, on the average, vary with a period of close to 11 years, for the 100-year period which the data covers.

Recently, it has been possible to reconstruct, in a crude way, the variation in solar activity for the last 2,000 years. The data obtained from ^{14}C studies of tree rings and reproduced in Fig. 2.9 show very low solar activity between 500–1100 A.D., reaching a deep minimum in about 650 A.D. Also noted are two strong minima occurring later: the Spører minimum at about 1450 A.D. and the Maunder minimum at about 1670 A.D.

The *Poetic Edda* was probably composed in the last part of the mediaeval minimum between 700 and 1100 A.D. From the relatively low solar activity during this period we can thus understand why the authors of the poems probably never mentioned the northern lights. Conflicting evidence, however, stems from the strong maximum at 1200 A.D. just at the time it is believed that *The King's Mirror* was written.

As already indicated, statements made in *The King's Mirror* may indicate that in about 1250 A.D. the northern light occurred more to the north in Norway than it does today. Our present knowledge shows that the northern light moves southward and covers a larger area in Scandinavia when solar activity increases. Based on the fact that solar activity was strong between 1100 and 1300 A.D., one could expect that the northern light at this time would have been fairly common in large parts of Norway. Since this apparently was not true, lack of northern light in Norway must have been due to reasons other than variations in solar activity.

The most likely candidate is variation in the Earth's magnetic field. By relying on world-wide archaeomagnetic measurements it has been inferred that the dipole axis has changed its position considerably, from 700 A.D. until the present. This is illustrated in Fig. 2.10, where it is seen that an anticlockwise rotation of the Northern Geomagnetic Pole has been predominant in the time span studied.

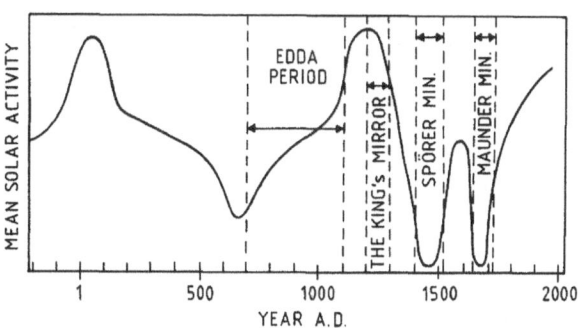

Fig. 2.9. Mean solar activity during the last 2,000 years. Also indicated are periods when the *Poetic Edda* and the *King's Mirror* were composed. The time periods for the Spører and Maunder minima are also shown for reference. The data are obtained by studies of ^{14}C-content in tree rings. (D. W. Hughes, Nature, Vol. 266, pp. 405–406, 1977)

Fig.2.10 The position of the geomagnetic pole in the Northern Hemisphere for the last 1,200 years, obtained by analysis of archaeomagnetic data. (N.Kawai et al., J.Geom and Geoel, Vol.19, pp.217–227, 1967)

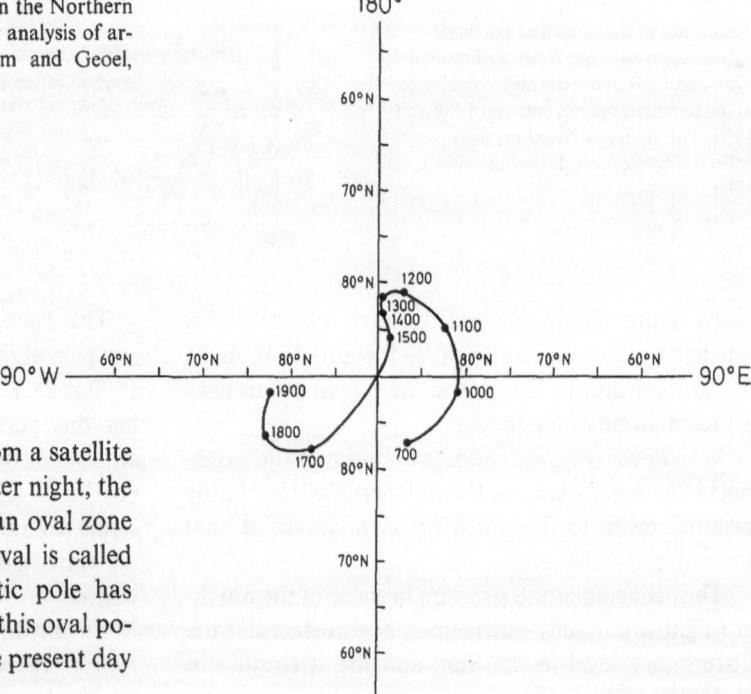

If one imagines a snapshot, taken from a satellite high above the polar cap on a dark winter night, the northern lights would be seen to form an oval zone encircling the geomagnetic pole: this oval is called The Auroral Oval. As the geomagnetic pole has changed its position from time to time, this oval position has also moved with respect to the present day auroral oval.

Even though it is realized that the depicted drift of the geomagnetic pole in Fig.2.10 is subject to large uncertainties these data have been used in Fig.2.11 to indicate a possible location of the nightside auroral oval for the years 1200 and 1970 A.D. An oval corresponding to a present time medium size oval has been used. For 1200 A.D. i.e. at about the time *The King's Mirror* was written the oval was probably far to the north of Norway. In spite of solar activity

Fig.2.11. The *left figure* illustrates how we today believe the average geographic distribution of the northern light was in the year 1200 A.D. The *right figure* shows the present position (1970).
(*Black dots* illustrate the position of the geomagnetic pole at the years indicated). About 1200 A.D., the northern light was probably well to the north of Norway, but was observable from Greenland

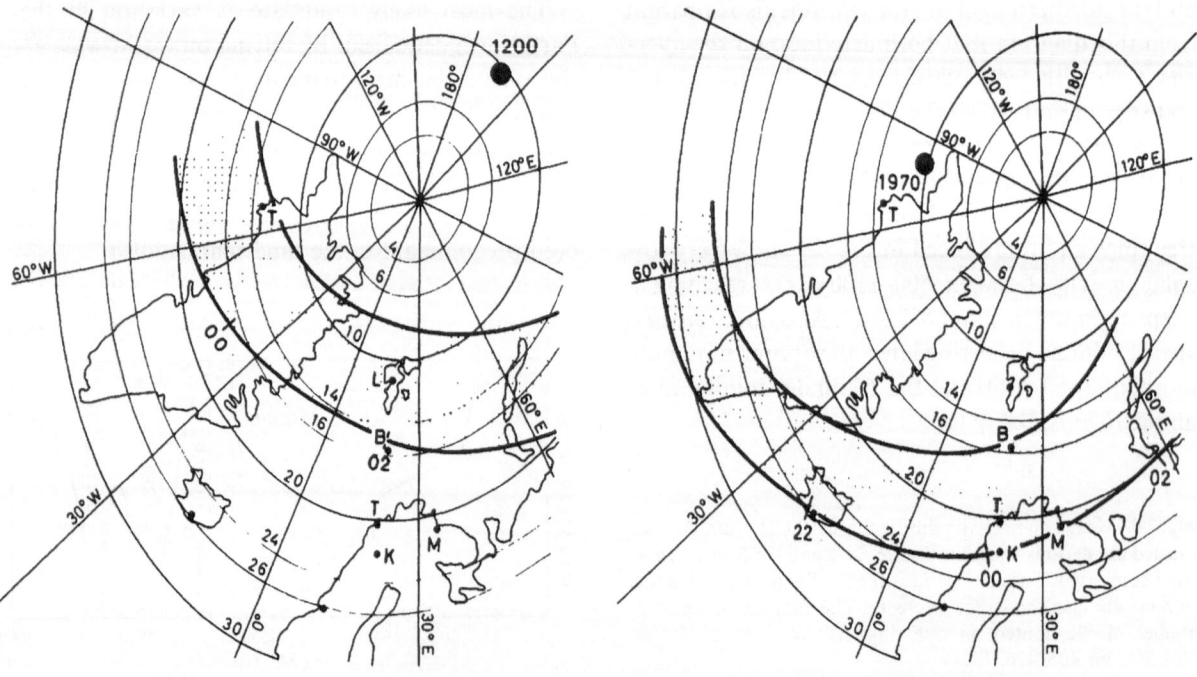

being so high at this time, only in extreme cases would it be possible to see the northern light from the middle of Norway.

With respect to Greenland, however, the oval shown here for the year 1200 A.D., also passes too far to the north for the northern light to be a common sight among the Norwegian settlement on Greenland. This settlement was at the southern end of the island, and from here the author of *The King's Mirror* implies that the northern light was seen.

In Fig. 2.11, the form of an oval as it would occur in today's geomagnetic field has been used. Paleomagnetic studies indicate that the form of the geomagnetic field has changed, at times being very different from the present dipole type magnetic field (Chap. 10.5).

When only modest deformations in the Earth's magnetic field are taken into account, it has been shown that the auroral oval is much deformed compared to the present-day oval. At the moment we know very little about the nature of these deformations, but it is not unlikely that they were relatively strong at the time *The King's Mirror* was written.

In about 700 A.D., i.e. before the first poem in *Edda* was composed, the nightside oval was probably into the southern part of Norway and moved progressively northward until about 1100 A.D., just after the last poem was probably composed. During this period, solar activity increased from low to modest. If one puts the *Edda* period between 1000 and 1100 A.D., as the latest research in this field indicates, one sees that based on this model, the oval just touches the northern tip of Norway. Since the authors of *Edda* probably lived in the southern part of Norway, Iceland and in England, the oval would be so far to the north that the northern light would be a rather rare phenomenon, particularly if one takes into account the low solar activity existing at this time.

Variation in the strength of the geomagnetic field will also influence the position of the northern light in such a way that the oval moves more poleward in periods of a strong field, if solar activity is constant. According to modern research the geomagnetic field, 1,000 years ago, was about 30% stronger than it is today and this would, together with low solar activity, place the auroral oval at very high latitudes during the time of the composition of *Edda*.

In conclusion, we therefore maintain that it is likely that low solar activity, together with a strong geomagnetic field, can explain the lack of references to northern lights in *Edda* (1000–1100 A.D.), and that deformations of the Earth's magnetic field probably must be inferred to explain the statements made about the northern light in *The King's Mirror* (ca. 1250 A.D.).

3 The Northern Light – A Source of Inspiration

3.1 The Northern Light and the Midnight Sun – Characteristic Manifestations of the Polar Sky

The midnight sun and the northern light perform their displays in the setting of the polar sky. The midnight sun goes through a monotonous daily, predictable motion guided by the Earth's rotation about its own axis and its revolution around the Sun. The oblique position of the Earth's rotational axis is the reason why the Sun, in a few hectic summer months, never dips below the horizon in latitudes north of the Arctic Circle.

The northern light, however, is a capricious phenomenon performing an ostentatious and unpredict-

able act – a kind of persona non grata event for those who would prefer to see the revelations of nature wrapped in strictly regular physical laws.

During the long, lively summer nights in the polar region, the midnight sun never sets and crowds of tourists are drawn to the polar summer. In the light summer the northern parts of Norway are filled with thousands of Southern Europeans and others who

Fig. 3.1. During an auroral display one has a chance to see the most fantastic forms and structures. The northern light can split up into long, thin rays, which can form displays by itself, or it can stay together and form coronas and draperies

want to admire the midnight sun, as if it were a sun different from the one to be seen in London or Lisbon.

In contrast to the midnight sun, performances of the northern light occur on clear and dark winter nights, and only for those hardy souls defying the cold polar night. But to these, the northern light is a dazzling spectacle, a kind of heavenly ballet dancer, exposing her pirouettes and arabesques in a manifold display of rich colours. Many Scandinavians have certainly shivered while watching the northern light, this heavenly dancer, and on ocassion thought they dimly perceived a faint accompaniment from an invisible orchestra.

One must admire the northern light, swinging its draperies in purple and scarlet colours in the setting of a deeply dark polar sky, itself sprinkled by twinkling stars and a perpetually brilliant Milky Way. It is all staged and designed by forces which one can only dimly perceive, and which the Nordic people have admired and feared for thousands of years.

Is this gigantic firework created as a solace to the polar people, who, being tired of the long polar nights, are reminded that the light will again return with its life-giving force?

The northern light is life-giving in its own way. It has affected the human imagination for generations before us, inspiring painters, authors, and other artists. Certainly it will continue to inspire humanity for as long as our descendants inhabit the polar regions.

Fig. 3.2. Hans Christian Andersen (1805–1875), the famous Danish fairy tale teller, let the northern light form a romantic setting in one of his stories

3.2 The Northern Light an Inspiration for Writers

The oldest Norwegian poem mentioning the northern light, except for probably a few lines in the *Poetic Edda*, is *The Northern Sea* by Simon Olaus Wolff (1796–1859). The third verse of this poem describes the cruelty of the Northern Sea in our own translation:

> The Sea is horrible whenever it is roaring,
> heavily against the nightly Coast of the North Pole!
> Swimming Icebergs, smash the Armor,
> thundering highly against the Breast of the Rocks.
> The Sepulchral Lamp of the Northern Light shines above
> Shroud of ancient, slumbering Lands;
> The Sea is not sleeping in Death chains tied,
> Strongly it breaks its Waves towards the Shore.

To the author of this poem, a parish pastor, the northern light was a dreadful experience leading his thoughts towards death and its unpleasant aura. His contemporary, Swedish author Esaias Tegnér (1782–1846), had a much more romantic view of this phenomenon. This can be seen from a few lines in the epic *Frithiof's Saga*. In the verse *Frithiof comes to King Ring* it says the following, again in our own English translation:

> The astonished queen's cheeks change colours as quickly,
> as red northern lights paint snowcovered fields,
> like two water lilies during a loud raging storm
> stand swinging in the bay, her bosom swelled.

The northern light unveils its identity by swiftly changing colours. In the same way Queen Ingeborg reveals some of her secrets with her blushing cheeks when she recognizes Frithiof.

Tegnér, however, also associated a sense of violence with the northern light, as in a few lines from the verse *Frithiof inherits from his father* in the same epic. Again in our own English translation:

> Viking left his sword to Thorsten, his son,
> and from Thorsten
> it went to Frithiof in heritage: when he drew it,
> the hall was lit up
> like a lightning rushed through there, or a
> northern light.

In the Norse literature one finds many heroes armed with shining swords. Here, Tegnér associates one of them with the northern light.

The famous Danish author Hans Christian Andersen (1805–1875) also mentioned the northern light in his fairy tales. In particular, the northern light forms a romantic setting in the story called *The Snow Queen*, where a little girl named Gerda is riding a reindeer to Lappland when the northern light is flaming in the sky:

a, b

c

Fig. 3.3 a–c. Many artists have tried to capture the elegance and beauty of a northern light with pen or pencil. These pictures were made by the French artist Bevalet who participated in the Recherche expedition to Bossekop in Finnmark during the winter 1838–1839. See also Fig. 11.1

",Fut! Fut!' it said in the sky. It looked as if it was sneezing red.
,It is my old northern lights!' the reindeer said ,look how they are shining!' and continued to run for days and nights."

It would be wrong to search in lyric poetry for a physical understanding of the features of the northern lights. Lyrical freedom permits poets alone to be somewhat frivolous in this respect. Jørgen Moe (1813–1882), another Norwegian minister in a poem from 1866, for example, expresses his thoughts about the northern light in the following way:

Reflections on the Northern Light

When the Greats visit the King
it is Glitter and joyful Parties
within the high Castle
Outside Torches flame
Candlelights arise – Framing the Rays
will then oust all Grieves.

See how the Heavenly Castle is blazing
The Castle of the King of Kings – encircled
by the glittering Flames.
Shooting upwards, spreading outwards
multicoloured Lights scattering
Stars are forming Garlands of Torches.

Dear, what is the matter
behind the lofty Fence of Rays?
Why this very splendid Feast?
Oh, a poor Sinner, selected
and on Jesus' Shoulders erected
is coming now as eternal Guest.

One finds Jørgen Moe's conception of the northern light in sharp contrast with Wolff's gloomy picture in a setting of death. But the brilliant rays and the torchlike flames which are so typical in a display of the northern light, have put Moe in a mood of fun and happiness.

3.3 Twentieth Century Norwegian Poems Touching on the Northern Light

In Scandinavia many authors have noted the northern light in one or more of their poems; either the northern light has been the main theme of their poems, or a setting, or merely an allegory. It would be beyond the scope of this book to mention them all; we will content ourselves by selecting a few Norwegian authors to illustrate how different writers deal with the same phenomenon.

The famous Nobel prizewinner Knut Hamsun (1859–1952) also associated the northern light with a heavenly feast, but in contrast to Jørgen Moe's welcome to the reception of a sinner in the realm of the

26

dead, Hamsun's occasion was a wedding among the stars. The poem, which is called *Snow*, is published in his collection of poems called *The Wild Choir* (1904) and from it this verse is translated:

> Flaming lights in the sky
> it is a lofty auroral night.
> A wedding party aloft
> where stars are loitering around
> and stars are forming brushes.
> The Moon is rising
> a god among her glittering goddesses.

In the same poem Hamsun also uses the northern light as a romantic setting for a rendevouz.

> It stands like rivers all white
> the quiet auroral night ...,
> Lonely steps in the snow are creaking,
> a young couple is having a meeting
> which nobody knows takes place.
> The braids so light
> and shining eyes are blinking in the auroral night –!

Jakob Sande (1906–1967) used the northern light as the background for a party he described in his racy

Fig.3.4. This illustrates some of the dark and gloomy impressions that Trygve Bjørgo describes in his poem. The figure is taken from the book *Naturkrefter* written by A. Paulsen in 1879

poem called *Hornelen*, from his anthology *The Cross and the Sledge Hammer* (1939). Here are quoted the last three verses of this poem:

> Northern lights blaze in the winter night
> above the dark gorges
> Quietly through the night a windharp whispers
> Mute death screams and giant's laughter rises
> up from the dark deep.
>
> Devils are dancing at midnight time
> the jew's harp cuts and shrieks.
> Roaring laughter and giant's play
> echoing in the night until cock-crow
> and the aurora twitches and shines.
>
> Here gather together from north and from east and west
> trolls and sorcerers.
> Here they arrange their holy midwinter feast
> where Judas is precentor and the Devil is priest
> while the northern lights rustle and burn.

Trygve Bjørgo is a current poet who perceives some of the same enchanting atmosphere beneath an auroral display. This can be sensed from his poem *Auroral Night* from his anthology *Darkness and Dawn* (1954). The first verse is quoted here:

> Deathcold clear night and moon
> rustling northern light fire.
> Steeple sharp glacier peaks
> stride towards the horizon

a

b

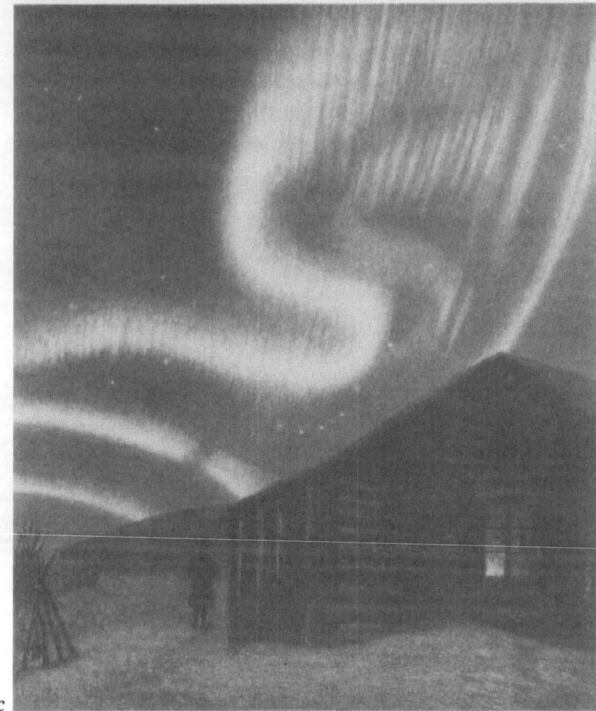

c

Fig. 3.5 a–c. These remarkably fine and realistic pictures of the northern light were made by the Danish painter Harald Viggo Greve Moltke (1871–1960) Moltke's aquarelles are some of the best-known artistic illustrations of the northern light

What a joy, what a living
what a blazing radiance!
Is it sunpeople swinging
in the playing dance?
Is it Milky Way sprites
jumping and leaping?
Is it elfins from their hides
in a sweeping swing?
I must stop and stand
I must look and stare.
It is a dusk of live embers
Then the sky is afire!
See the surf horses beating
and the clearing glinting
Above the highest peaks.
Oh, there it is extinguished
No, now it is brushing forward
with a rustling fire.
And it is quivering and shaking
like a fringed shawl.

in the far north of the giant's height
where desolation and ice prevail.
The vault with stars is scattered
pale in the horizon glint.

Both Sande and Bjørgo are inspired by the rustling motion of the northern light, but Lars Eskeland (1880–1964) more than anyone else has managed to mediate the speed and rapid variability in the northern lights in his poem called *Northern light* in the collection *Upwards* (1921). The introduction to his poem is recited here:

As with so many others, Eskeland's mind is reminded of dancing when he sees the northern light. To him the dance can be performed by anyone from people to elves to sprites. Eskeland also, however, plays with the idea of a sound associated with the

28

phenomenon. This is a very common metaphor used by several poets and is, of course, related to the fact that a great many people think that the northern light is accompanied by a rustling sound.

The poem ends with a few thoughts about the forces behind the phenomenon, and Eskeland associates the northern light with flames from the hearth. A hearth, in his mind, is where life is formed and chiseled into shape. In Eskeland's poem the northern light has a double role: simultaneously with inspiring his thoughts towards the infinity of space and the power behind it, he associates the northern light with the contrast between perishability and eternity of life.

There are other poets who have depicted this ability of the northern light to lead the mind towards infinite space. Jan Magnus Bruheim touched upon this in his poem *In an auroral night*, from his anthology *The Letter to Love* (1977). The first verse reads as follows:

> On the crackling cold sky above the mountains,
> flames are playing,
> It is a longing night.
> It is an auroral night.
> Uneasy rambling searchlight inwards – inwards,
> towards the endless, unprobed space.
> It shoots sharp arrows of flames
> deeper and deeper inwards.
> Penetrating towards the heart of the
> sky, but pulling them back without hurting.
> Glimmering and alive, fingering and
> playing a space symphony. One seems as if
> hearing rustling and plinking of stringed instruments.
> The light arrows hunt and search deeper
> inwards in space. Without finding, without receiving.

Bruheim also thinks he perceives a rustling sound while watching a northern light display, or a sound like music emerging from a stringed instrument. The sound of the northern light has also been noted by the author Harald U. Sverdrup (1929–) in his poem *Norwegian Landscape or Spring Feelings*, from his collection of poems called *Perpetual Building Babel* (1949). In this poem we find the line:

> The frozen auroral bonfire crackles towards zenith.

Another example can be found in a poem *Nights in Northern Norway* in an anthology by Villa Trap (Wahl) (1909). A line in this poem is as follows:

> The northern light, strays, whistles, leaps,
> clicking clicks like castanets.

The northern light may also appear feather-light and transparent, as Rolf Jacobsen touches upon in

his poem *Wash and Dry Inc*, in the collection *Breathing Exercise* (1975). The second verse reads as follows:

> During the night the northern light comes
> with white chains
> and hangs it up to dry
> in the starwind.
> Bluewhite and clean, but thin,
> Shirts and long slips
> – almost like the angels' garment.

There are about 50 Norwegian poets who have mentioned the northern light in one or more of their poems, and a similar number of authors can be found in the other Nordic countries. One could continue like this and recite a large amount of them, but here we will quote only one more. One cannot avoid mentioning, even in a short review like this, the poem *Northern Light*, by Theodor Caspari (1853–1948), in his collection of poems called *Picture of the Times* from 1883. The poem was undoubtedly written as a result of the great interest created in the northern light in Norway by the organization of *The First International Polar Year*, 1882–1883. As a result of this, Norwegian scientists constructed two small observatories in the northern part of Norway (cf. Chap. 11). The poem was later published as a tribute to the Norwegian pioneers in auroral science. Later on, Willy Stoffregen, a well-known Swedish auroral scientist, composed a melody to this poem, the notes to which are reprinted in Fig. 3.6. Caspari's poem follows here in its entirety:

Northern Light

> Sparkling Arc,
> playing Flame,
> Flickering garland around the mistful Pole.
> Iceful blaze
> pictureless Frame
> you are to me, Aurora, a symbol of life.
>
> The thoughts are created,
> strengthened and scattered,
> twined and twisted in Generations by Hands.
> The sparks are meeting,
> agreeing and linking
> Generation after Generation as flickering Bands. –
>
> Twinkling Stars,
> rolling Spheres
> circling aloft in light Harmony.
> Thoughtful Brains
> Modest Heads
> Ignite down here a Firework.
>
> See, how it is shining
> Lightning Lances,
> Silver Tiara of the Illumination's Treasure.
> The Generations are freezing
> rejoicing and dancing
> restless about in the eternal Night. –

Rays after Rays
gleaming, certainly
emerging from the Hall of Research –
Needles after Needles
icing, sharp
cheering the seething Life in Crystal.

Pictureless Frame
little you benefit, –
Research name for a sick Eskulap
when behind your Flame
Mystery behind Mystery
is forming a bluish Chaos.

We notice that Caspari explains the northern light with the help of ice needles. At the time Caspari republished his poem in 1921, the Norwegian physicist Lars Vegard (Chap. 7.4) was working on a theory of the northern light, based on spectroscopic ob-servations, in which he claimed that the light was emitted from frozen nitrogen. It is likely that Caspari, who had written his poem in 1883, felt that he had been in advance of his time and with the publication of Vegard's work, found an opportunity to republish it almost 40 years later.

3.4 Colourful Descriptions of the Northern Lights

One of the leaders of the Austrian/Hungarian polar expedition to Franz-Josef's land in 1872–1874, captain Karl Weyprecht (1838–1881), named the groups of islands after his famous patron. He was also the principal organizer of the First International Polar Year (1882–1883) (cf. Chap. 11). Unfortunately,

Fig. 3.6. The original notes of Willy Stoffregen's (see Chap. 11.5) composition to the poem the *Northern Light* by Theodor Caspari. The text is given in its original language.

ig. 3.7   *Fig. 3.8*

he died before he could see his great enterprise carried through. The following is a free translation of Weyprecht's feelings when he watched a beautiful northern light.

"From the center issues a sea of flames: is that sea red, white or green? Who can say? It is all three colours at the same moment. The rays reach almost to the horizon: the whole sky is in flames. Nature displays before us such an exhibition of fireworks that it transcends the power of imagination. Involuntarily we listen: such a spectacle must, we think, be accompanied by sound. But unbroken stillness prevails; not the least sound strikes the ear. Once more it becomes clear over the ice, and the whole phenomenon has disappeared with the same inconceivable rapidity with which it began, and gloomy night has again stretched her dark veil over everything. This was the northern light of the coming storm – the northern light in its fullest splendour. No pencil can draw it, no colours can paint it, and no words can describe it in all its magnificence."

In spite of this last statement, Fridtjof Nansen (1861–1930), the world famous polar explorer made several attempts both to draw and describe it with his pencil. In particular, in his book *Farthest North* from 1897, he gave several accounts of the phenomenon representing some of the most successful descriptions of northern light in Norwegian literature:

"I went on deck this evening in rather a gloomy frame of mind, but was nailed to the spot the moment I got outside. There is the supernatural for you – the northern lights flashing in matchless power and beauty over the sky in all the colours of the rainbow. Seldom if ever have I seen the colours so brilliant. The prevailing one at first was yellow, but that gradually flickered

Fig. 3.7. Karl Weyprecht (1838–1881) Austrian naval officer who initiated the idea of the First International Polar Year which was carried out after his death (1882–1883). On his expedition to the Arctic he observed and described several northern light displays

Fig. 3.8. Fridtjof Nansen (1861–1930), the world-famous explorer, was a man with many fields of endeavour – he was an established zoologist, a reputable diplomat, and a skilled artist. This is a self-portrait from 1930

over into green, and then a sparkling ruby-red began to show at the bottom of the rays in the under side of the arch, soon spreading over the whole arch. And now from the faraway western horizon a fiery serpent writhed itself up over the sky, shining brighter and brighter as it came. It split into three, all brilliantly glittering. Then the colour changed. The serpent to the south turned almost ruby-red, with spots of yellow; the one in the middle, yellow; and the one to the north, greenish-white. Sheaves of rays swept along the sides of the serpents, driven through the ether-like waves before a storm wind. They sway backwards and forwards, now strong, now fainter again. The serpents reached and passed the zenith. Though I was thinly dressed and shivering cold, I could not tear myself away till the spectacle was over, and only a faintly glowing fiery serpent near the western horizon showed where it had begun. When I came on deck later the masses of light had passed northwards, and spread themselves in complete arches over the northern sky. If one wants to read mystic meanings into the phenomenon of nature, here, surely is the opportunity".

Also in the same chapter, one finds the following poetic description of another auroral display:

31

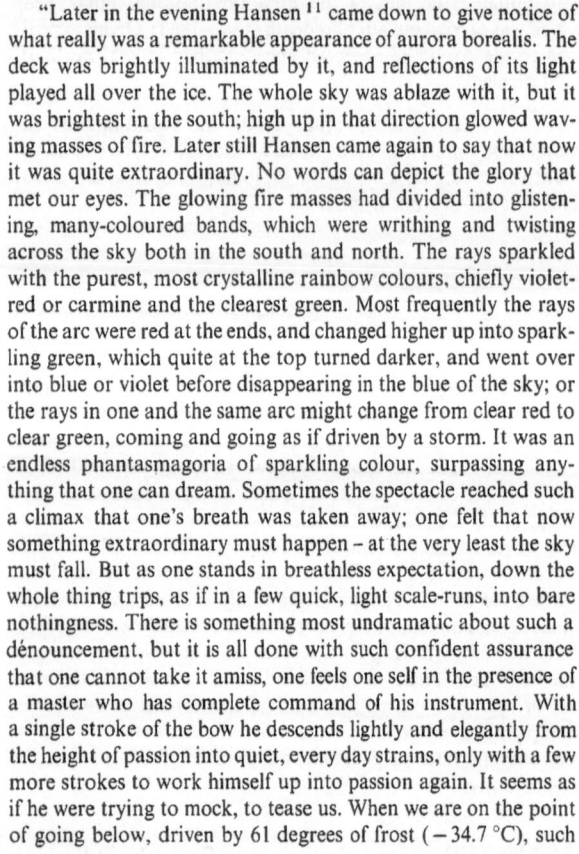

Fig. 3.9 a–c. Fridtjof Nansen made several drawings of the northern light as illustrations for his many books; these are vignettes to *The Northern Mists* (Nord i Tåkeheimen) (1910–1911)

"Later in the evening Hansen [11] came down to give notice of what really was a remarkable appearance of aurora borealis. The deck was brightly illuminated by it, and reflections of its light played all over the ice. The whole sky was ablaze with it, but it was brightest in the south; high up in that direction glowed waving masses of fire. Later still Hansen came again to say that now it was quite extraordinary. No words can depict the glory that met our eyes. The glowing fire masses had divided into glistening, many-coloured bands, which were writhing and twisting across the sky both in the south and north. The rays sparkled with the purest, most crystalline rainbow colours, chiefly violet-red or carmine and the clearest green. Most frequently the rays of the arc were red at the ends, and changed higher up into sparkling green, which quite at the top turned darker, and went over into blue or violet before disappearing in the blue of the sky; or the rays in one and the same arc might change from clear red to clear green, coming and going as if driven by a storm. It was an endless phantasmagoria of sparkling colour, surpassing anything that one can dream. Sometimes the spectacle reached such a climax that one's breath was taken away; one felt that now something extraordinary must happen – at the very least the sky must fall. But as one stands in breathless expectation, down the whole thing trips, as if in a few quick, light scale-runs, into bare nothingness. There is something most undramatic about such a dénouncement, but it is all done with such confident assurance that one cannot take it amiss, one feels one self in the presence of a master who has complete command of his instrument. With a single stroke of the bow he descends lightly and elegantly from the height of passion into quiet, every day strains, only with a few more strokes to work himself up into passion again. It seems as if he were trying to mock, to tease us. When we are on the point of going below, driven by 61 degrees of frost (− 34.7 °C), such

[11] One of the crew-members

mangificent tones again vibrated over the strings that we stay, until noses and ears are frozen. For a finale, there is a wild display of fireworks in every tint of flame – such a conflagration that one expects every minute to have it down in the ice, because there is not room for it in the sky. But I hold out no longer. Thinly dressed, without a proper cap, and without gloves, I have no feeling left in body or limbs, and I crawl away below".

And finally Nansen describes a beautiful corona in the following manner:

"A lovely aurora this evening (11.30). A brilliant corona encircled the zenith with a wreath of streamers in several layers, one outside the other; then larger and smaller sheaves of streamers spread over the sky, especially low down towards S.W. and E.S.E. All of them, however, tended upwards to the corona which shone like a halo. I stood watching it a long while. Every now and then I could discern a dark patch in its middle, at the point where all the rays converged. It lay a little south of the Pole Star, and approached Cassiopeia in the position it then occupied. But the halo kept smouldering and shifting just as if a gale in the upper strata of the atmosphere were playing the bellows to it. Presently fresh streamers shot out of the darkness outside the inner halo, followed by other bright shafts of light in a still wider circle, and meanwhile the dark space in the middle was clearly visible; at other times it was entirely covered with masses of light.

Then it appeared as if the storm abated, and the whole turned pale, and glowed with a faint whitish hue for a little while, only to shoot wildly up once more to begin the same dance over again. Then the entire mass of light around the corona began to rock to and fro in large waves over the zenith and the dark central point, whereupon the gale seemed to increase and whirl the streamers into an inextricable tangle, till they merged into a luminous vapour, that enveloped the corona and drowned it in a deluge of light, so that neither it, nor the streamers, nor the dark centre could be seen – nothing, in fact, but a chaos of shining mist. Again it became paler, and I went below. At midnight there was hardly anything of the aurora to be seen".

A little-known English clergyman, Richard Carter Smith (1802–1864), made a long journey in Europe, and in the summer of 1838, he visited Norway. In the evening of August 23 of this year he observed a beautiful northern light at Frogner, in Oslo. He gave the following description of it in a manuscript under the title of *Travel in Norway 1838*[12].

"We observed it when it had risen about 20 degrees above the horizon. Gradually, the aurora came closer, and it became more and more radiant until Ursa Major was in the middle of a broad shining band. As it moved up toward the Polar Star it started to emit light rays down toward the northern horizon. Soon thereafter the uttermost points on the huge arc became clearer, and suddenly they emitted light beams straight up to the zenith, until they disappeared to make a portal above our heads. Then the beams bent more southward, and finally they covered close to two-thirds of the sky. The light radiated incessantly from horizon to zenith and changed continually in intensity. Anyhow it had a uniform pale-blue tint all over except for a couple of places in the northwest where the light beams had a weak reddish tint. It looked as if we were standing underneath a shining firedome, and the gleam of light was so strong that it made a clear shadow. About an hour later the aurora became weaker. Its beauty cannot be described by words, but the Norwegians pretended not to be too excited about it".

One of the first foreigners to travel in Norway and leave a description of the northern light was the Italian priest and explorer Francesco Negri (1623–1698). In 1663, he left his home town of Ravenna, where he was a priest, and travelled to Northern Europe. He came to Norway in 1664 visited Finnmark and North Cape, and returned to Italy in 1666. While in Norway he visited the chancellor, Ove Bjelke, from whom he received recommendations for his journeys and to whom he reported by letters. His account of the journeys was published posthumously by the title *Viaggio settentrionale fatto e descritto dal molto Rev^de Sig^r D. Francesco Negri da Ravenna*. From an Italian to Norwegian translation, we translate into English his description of the northern lights:

"I have seen even one more phenomenon, which is very common in this cold zone, which I have never seen before, and which has probably never been observed in Italy in my lifetime. It is during the long winter nights, that one can take advantage of this magnificent sight in its manifold splendour. Once I saw it like a long cloud starting about three degrees above the horizon and stretched all the way towards zenith and to the same distance on the other side of the sky. It was all clear and transparent such that it even threw a weak light onto the ground. It changed between all possible forms, once an arc, once a brilliant corona, once a snake. Sometimes the shining mass split up and moved in opposite directions, and later merged together again to form new rays, which stretched into endlessness, faded and disappeared. Once, I remember, I saw a large number of those being in length about the size of a man (as it looked), one after the other in a long row, almost like soldiers defiling at high speed, and all the rays gathered then into a large pencil of rays which later seemed to increase. Thus I always felt that the figures resembled known statues and therefore was guided by my fanciful ideas. The forms change indefinitely and are so beautiful and brilliant, that they by no means create apprehension, but offer a vision, which to me appears as some of the most delicious one can see in the world. This phenomenon is very rare in the temperate zone, where one calls them *capre saltani* (flying goats), *tizzoni ardenti* (flaming trees) etc. I do not, however, believe that this is the same thing since these are caused by emanation from the ground".

And finally, we include another description of the most beautiful of all auroral displays, the corona. The description was originally written in Norwegian by Sophus Tromholt (1851–1896) (Chap. 6.5) and here follows our translation into English:

"High up, vertically above our heads, new bands crossing the sky are formed, which move rapidly towards the south and disperse. Now the light masses have passed zenith, and the rays shoot towards one point high up in the southern sky, in the east and west the rays move still further southwards. Then a strange sight appears. The whole sky, in all directions, is covered with raybundles, and they all shoot towards that point (the magnetic zenith) and so transform the vault of heaven into a mighty dome of flames, its beauty no word is able to express and no brush is able to paint. All the brilliant shades of colour, which together compose the seven coloured rainbow, are here meeting to decorate the glorious vault of light; there are the green of the emerald, the red of the ruby, the blue of the sapphire. Here a body of yellow-green colours is frolicking in a merry game; here mighty columns are stretching up as if they are supporting the vault of the proud building, there the sky looks as if it is covered with a curtain of a deep red, transparent substance, and behind the curtain white rays emerge and shine through. That is the corona of the northern light. The human eye will certainly never see a more beautiful vision; anyone who has never seen it cannot conceive of this wonderful display, which defies any description".

---

[12] Richard Carter Smith: "Reise i Norge 1838", translated by Johnny Johnsen, Universitetsforlaget, Oslo, 1976

4 Accounts of Northern Lights in Scandinavia –
From the Viking Era to the Renaissance

4.1 Old Known Auroral Records

The oldest written records by Norwegians concerning the northern light go back to around 1000 A.D. In order to delve further back into the history of this subject, one needs to study the literature and art from the Mediterranean area and from China. Since the northern light mainly occurs in the Polar regions, the ancient philosophers living in areas around the Mediterranean and in China had very little chance to see even one or at best a very few northern lights during their lifetime because of their geographic locations. It is not possible to determine accurately when the first northern light, as seen by man, along with his impression of it, was recorded in some manner for posterity. Some people think, and perhaps for good reasons, that a great deal of the very ancient engravings which have been found in several grottos along the Mediterranean Sea are in fact pictorial representations of the northern light.

It is highly probable that the following, taken verbatim from Ezekiel of the Old Testament (Revised Standard Edition) is in fact a description of a very active display of a northern light. We quote here sections from the first chapter of the Book of Ezekiel (written around 593 B.C.).

"As I looked, behold, a stormy wind came out of the north, and a great cloud with brightness round about it, and fire flashing forth continually and in the midst of the fire, as it were gleaming bronze. And from the midst of it came the likeness of four living creatures. ... but each had had four faces and each of them had four wings. Their legs were straight, and the soles of their feet were like the sole of a calf's foot; and they sparkled like burning bronze. ... over the heads of the living creatures there was the likeness of a firmament, shining like crystal ... I saw as it were gleaming bronze, like the appearance of fire enclosed round about ... like the appearance of the bow that is in the cloud on the day of rain, so was the appearance of the brightness round about".

This is not a bad poetic description of an active northern light and not so different from the way we might describe it today. There are several other passages in the Old Testament which one might argue are also descriptions of the northern light, e.g., Jeremiah

(1:13) and Zacchariah (1:8) but none of them are quite so convincingly descriptive of the northern light as the above account given by Ezekiel.

In China there is a relatively large amount of material on the northern light going back to around 200 B.C. These records show at least 170 auroral occurrences between 100 and 1000 A.D. along with many fantastic illustrations. The first record, to the best of our knowledge, goes back to 208 B.C. One brief segment from a Chinese record of an auroral event, observed 2,000 years ago, is roughly as follows: During the night luminous clouds were seen, gold and white, with long streamers which lit up the hills. Some say that it is Heaven's Sword, but others think that it is a deep hole, with a large blazing fire in the sky.

4.2 How Ancient Philosophers Regarded the Northern Light

In order to understand how familiar people of the Northern hemisphere were with the northern light compared to their contemporaries in the more scholarly cultures of Southern Europe, we shall first give a short resumé of the different theories which existed before the Viking Era.

One of the first attempting to explain an enhancement or unusual brightness of the sky was Anaximedes (570–526 B.C.). He believed that when a volatile material or vapour (i.e., a gas) was stored in the cloud and then slowly became mixed in the assumed boiler or heater, that an increase in brightness should occur.

Anaxagoras (500–428 B.C.) interestingly enough was charged with being godless because he contended that the Sun being a glowing, red-hot body. He proposed that a fiery vapour poured from the uppermost parts of the heavens down into the clouds. Here, he said, the vapour accumulated to the point where a fire blazed up; if the fire moved upwards it was accompanied by many different lights and if the

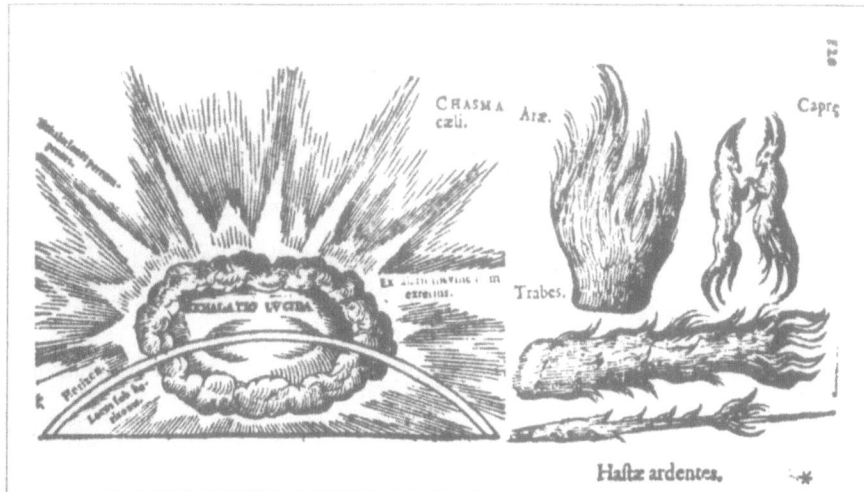

Fig. 4.1. Aristotle used many different names for the northern light and some of these are illustrated above. This illustration was made by Gemma in the year 1575 and represents one of the oldest illustrations of the northern light in Europe. One of the most fanciful names that Aristotle used was "jumping goats" which is not a bad poetic picture of auroral rays. This can be clearly seen in the enlargement to the right. The illustration to the left is taken from F. Link, Visitas in Astronomy, 1967. The enlargement to the right is made by the authors

fire moved downwards it created strokes of lightning. It is not being too speculative to assume that at least some of the lights as described here by Anaxagoras were auroral displays.

In Western European culture Aristotle (384–322 B.C.) is usually credited with being the first to discuss the northern light in a sound scientific manner. In a letter to Alexander the Great, Aristotle wrote about different light phenomena occurring in the night sky. Some of these lights, he stated, shot out at great speed while others remain stationary. Some flicker up and die out almost instantly while others remain on for a longer time. These light phenomena can present themselves in different figures like light torches, small rays, round vessels or barrels, and some appear as very abysmal or chasmatic.

A more scholarly work by Aristotle entitled *Meteorologia* is a textbook describing natural phenomena and meteors occurring in the atmosphere. Included among these are clouds, rainbows, halos around the Sun and Moon, lightning and thunder, hurricanes, and different kinds of light in the sky. In this treatise, he mentions a kind of light which streaks far out into the atmosphere and reminds him of flames from burning grass. There are times when the light spreads out in breadth while simultaneously scintil-

lating with luminous rays, and these he calls "jumping goats" whereas when the rays are absent he calls the light a pure and simple fire (Fig. 4.1).

In modern literature when the work on the northern light by Aristotle is referred to, one usually finds that he uses the general word chasmata. This could be very misleading because Aristotle had at his disposal a whole gamut of names and descriptions, each used according to what the northern light revealed to him. The most remarkable expression in this regard as used by Aristotle is "jumping goats" which at first glance might appear confusing indeed, but when we read ancient auroral accounts it is easy to imagine what he was trying to describe.

In order to understand Aristotle's explanation of his so-called "meteors", we must delve a bit into the picture of the world as the ancient Greeks perceived it during Aristotle's time.

During that time it was believed that the space around the Earth is filled with different sorts of vapour which could evaporate from the Earth. The moist vapour rising from the sea and lakes is the roughest and the heaviest of the vapours. From earthly materials come dry vapours, which are lighter in weight, are more fleeting and therefore rise to the highest elevations in the atmosphere. It was believed that both of these types of vapour fill the space all the way to the Moon and this space was called the atmosphere. All light phenomena occurring in the atmosphere were labelled with the joint word "meteors" which according to Aristotle results from an unstable mixture of four main elements; fire, air, water and soil. Because of the existence of this unstable mixture, the atmosphere is constantly undergoing changes in wind and weather.

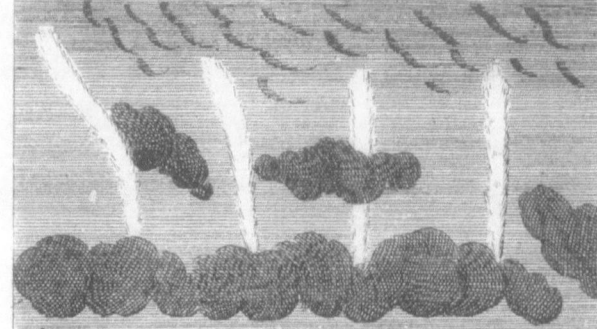

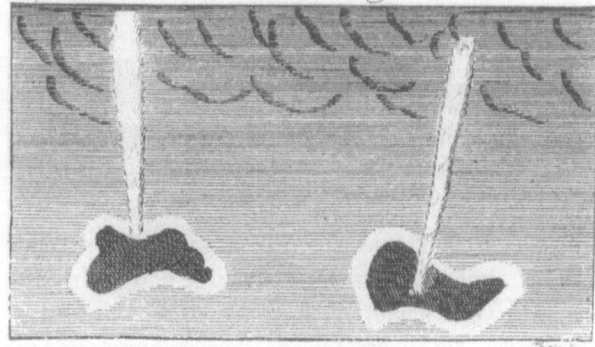

1. They are due to the fact that the air is boiling and starts to move,
2. strong winds create such phenomena,
3. they are simply due to the heat in the upper atmosphere,
4. air containing the fire is self-propelled into the sky,
5. because of their rapid movements stars produce sufficient heat to enhance the material from which these light phenomena are created.

Seneca also thought that different colours of light are due to the fact that material of varying composition are set afire.

The Roman official Plinius (23–79 A.D.) with his work *The History of Nature* added to the confusion for many further years. From the twelve different light phenomena which he lists, the following six descriptions might have been dealing with northern light:

1. A light from above surrounded by beams, resembling blood red hair.
2. Flames dashing rapidly backwards and forwards like shot arrows.
3. A corneous lightning figure.
4. Burning torches.
5. Light beams running about each other with great speed.
6. Lightning that resembles a sword.

It is believed that Plinius in his use of the word "comets" was referring to all these different brightenings of the night sky. After Plinius it was therefore common practice to give all light phenomena in the atmosphere this same name.

4.3 The King's Mirror Gives the Vikings' Theories About the Northern Light

In the Middle Ages the northern light was interpreted as containing frightful warnings about war and misfortune (Chap. 1). In view of this, it is rather amazing to read the prosaic description of the northern light in *The King's Mirror* from about

Among the "meteors" there are some which glow and burn, and according to Aristotle they are created in the following way: The lightest and driest of the gases rise to the highest elevations in the atmosphere and are there freed of all humidity. In this very dry condition they can therefore easily be ignited when in close contact with fire or heat. Continuing with Aristotle, fire is the lightest of the four elements and it moves to the top of the atmosphere. When the fire contacts the driest gases, they become ignited, causing the various light phenomena to occur.

If the dry gases collect in large quantities a rather large flash of fire is needed to burn them up – such is the manner in which long-lasting lights are created. If, on the other hand, the accumulation of dry gases is small, only short-lived light phenomena arise. The various figures and forms which Aristotle mentions in connection with the different light phenomena are explained by the fact that dry gases gather in different formations.

The stoic Seneca (5 B.C.–65 A.D.) followed along the same line of thinking as Aristotle, but modified somewhat his theory as to how lightning in air was formed. He used the following ideas to explain how such enhanced lights in the night sky are created:

1250 A.D. (Chap. 2.2). In the book which was written as a dialogue between father and son, the father says the following, in addition to what is already cited in Chap. 2.2, concerning the northern light (extract from *The King's Mirror*).

"However, it is true of that subject as of many others of which we have no sure knowledge, that thoughtful men will form opinions and conjectures about it and will make such guesses as seem reasonable and likely to be true. But these northern lights have this peculiar nature, that the darker the night is, the brighter they seem, and they always appear at night but never by day, – most frequently in the densest darkness and rarely by moonlight. In appearance they resemble a vast flame of fire viewed from a great distance. It also looks as if sharp points were shot from this flame up into the sky, these are of uneven height and in constant motion, now one, now another darting highest; and the light appears to blaze like a living flame. While these rays are at their highest and brightest, they give forth so much light that people out of doors can easily find their way about and can even go hunting, if need be. Where people sit in houses that have windows, it is so light inside that all within the room can see each other's faces. The light is very changeable. Sometimes it appears to grow dim, as if a black smoke or a dark fog were blown up among the rays; and then it looks very much as if the light were overcome by this smoke and about to be quenched. But as soon as the smoke begins to grow thinner, the light begins to brighten again; and it happens at times that people think they see large sparks shooting out of it as from glowing iron which has just been taken from the forge. But as night declines and day approaches, the light begins to fade; and when daylight appears, it seems to vanish entirely.

The men who have thought about and discussed these lights have guessed at three sources, one of which, it seems, ought to be the true one. Some hold that fire circles about the ocean and all the bodies of water that stream about on the outer side of the globe; and since Greenland lies on the outermost edge of the Earth to the north, they think it possible that these lights shine forth from the fires that encircle the outer ocean. Others have suggested that during the hours of night, when the Sun's course is beneath the Earth, an occasional gleam of its light may shoot up into the sky; for they insist that Greenland lies so far out on the Earth's edge that the curved surface which shuts out the sunlight must be less prominent there. But there are still others who believe (and it seems to me not unlikely) that the frost and the glaciers have become so powerful there that they are able to radiate forth these flames. I know nothing further that has been conjectured on this subject, only these three theories that I have presented; as to their correctness I do not decide, though the last mentioned looks quite plausible to me."

The idea that the Arctic snow and ice absorb so much light from the midnight Sun during summer time and radiate it to the northern light in winter time, is new in *The King's Mirror*, and this original work exerted an influence on many authors from that time on.

In Southern Europe the northern light created a great deal of interest in the beginning of the 1600's. In 1618 three comets came into view which led to a severe debate about the origin of comets, and here Galileo Galilei (1564–1642) was deeply in-

volved. In his discussions, Galilei mentioned the aurora by the Latin name "*boreale aurora*" or "the northern light of dawn". This description is rather misleading to those familiar with the northern light and associate it with its dominating yellow green colour. It is, however, a name that was undoubtedly used in regard to the phenomenon because the northern light, observed from the latitude of Rome, is as a rule seen to be reddish.

On 13 September 1621, a northern light appeared in France which the astronomer Pierre Gasendi (1592–1655) in his book *Physics* from 1649 denoted with the Latin name "*aurora borealis*". In international literature aurora borealis has become the name of what actually should be called *lumine boreali* or *lumine septentrionali*, the "northern light", as the author of *The King's Mirror* so strikingly christened the phenomenon.

4.4 An Ancient Arabian Description of the Northern Light in Scandinavia

Except for *The King's Mirror* and the *Edda* poems (Chap. 2) there are very few written sources which give knowledge of the northern light in Scandinavia before, say, 1500 A.D. There is, however, an interesting book written by the Arab Samsaddin al-Ansari as-Suft ad-Dimasqi who died in Rabwa in Syria in the year 1327 A.D. The book, which is a cosmographical work, carries the Arabian title *Nuhabat ad-dahr fi "aga" ib al-barr wa-l-bahr* or translated freely into English *The choice of time of strange beings on land and in sea*. Apparently ad-Dimasqi must have met some people from the far north who told him about the northern light, since he writes the following concerning this matter:

"In the neighbourhood of the Ice Sea (The Baltic Sea) at a distance of twenty days' journey and to the west of that, and to the north of the lands of Kilabiya, there is a great lake called the Shining Sea. The shores of it are inhabited by a people called As-Saqaliba. Sometimes during the night one sees light rays radiating, as of usual light without any fire or any shining material body being present, as it is for the light of the stars or the radiation from a fire".

Another source of interest in this regard is a few sheets from the ruins of Palmyra in Syria. It has been maintained that the northern light is described on these sheets and that the description is based on reports brought to Palmyra by people living in the area around the White Sea. How old these sheets are is not well known, but since Palmyra was destroyed in the year 273 A.D. they are probably older.

4.5 Olaus Magnus' Reference to the Greek Scholars

In 1550 a new writer on the northern light emerged in Scandinavia. His name was Olaus Magnus (1490–1558) and he wrote a large and well-known book entitled: *Scandinavia's History of the Northern People* or by its Latin title *Historia de Gentibus Septentrionalibus*. In this book, covering a variety of subjects, one finds a few lines devoted to what he calls comets and which he says almost always appear in the north. It is quite possible that these comets were in fact northern lights. His explanation of the phenomena is based upon the view held by the old Greek scholars that different vapours arose from the Earth. As these vapours collect in the atmosphere, a peculiar difference develops between them. They fight each other until they reach the north where the atmosphere becomes densest and here arise terrifying comets which predict dire situations such as hunger by starvation, war, and violent hurricanes. Because of the uninviting appearance of the comets people were startled and frightened, but Olaus Magnus was strongly in doubt that comets were omens of war and misery.

Magnus also describes some sharp and continuous lightning which appears on clear nights in September, and he says that this lightning is actually more frightening than physically harmful to the

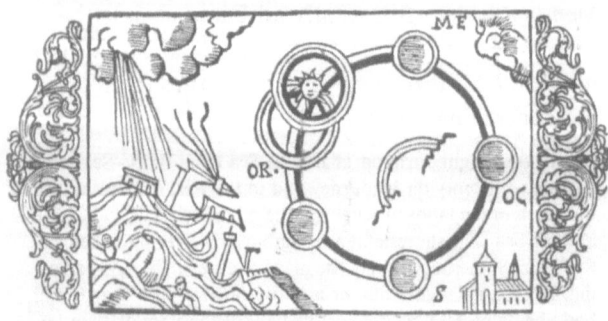

Fig. 4.3. In the book by Olaus Magnus (1490–1558) *Historia de Gentibus Septentrionalibus* there are several vignettes illustrating different natural light phenomena. In the text following this vignette Magnus describes a frightful comet appearing in the north as being an omen of famines, wars, and hurricanes. The monster shown here in the upper right corner probably illustrates this frightful comet. Both the position of the phenomenon and omens associated with it make it reasonable to think that it actually is an illustration of the northern light

viewer. The cloud from which the lightning originates does not contain enough material to create a stroke of lightning. This idea reminds us of that held by Anaxagoras almost 2,000 years earlier and it is quite possible that Magnus was aware of Anaxagoras' work. This was, according to Olaus Magnus, the learned scholar's explanation of these light phenomena which must have been northern lights.

4.6 The Famous Astronomer Tycho Brahe Observed the Position of Auroral Coronas with Great Accuracy

The last half of the 15th and first half of the 16th century were periods of low auroral and solar activity. It was first noticed in historical data on sunspots, and the period from 1460 to 1550 is now called the Spører minimum with reference to the discoverer of this period of low solar activity. From this period very few, if any, data exist in Scandinavia concerning auroral observations. Immediately after this period, however, we know of a collection of meteorological notes from 1565–1572 by a Danish abbot called Morten Pedersen (Alban) (1537–1594). The notes have been written in a copy of the so-called Paul Ebers Calendarium in Wittenberg in 1571 and contain a few dated descriptions of the northern light.

A Swedish author Georgius Olaus stated in his writings from 1588 that it was more common to see the northern light between 1560 and 1580 than ever before, in good agreement with a long period of auroral quietness in the Spører Minimum.

In 1576 King Frederick of Denmark gave the Island of Ven to the famous Danish astronomer Tycho Brahe (1546–1601). Here Brahe established his observatory, Uraniborg, and started a school of modern astronomy. Brahe had a well-known reputation as a very careful and methodical observer which Johann Kepler (1571–1630) so clearly discovered when he received Brahe's data after his death, and from these promulgated the three laws of planetary motion.

Brahe also observed several northern lights in his lifetime. In his *Meteorological Journal* he used the name Chasmata from Aristotle to describe this phenomenon. Sometimes he mentions the auroral corona with a very precise description of its position among the stars. At other times, both the azimuth and the elevation of the auroral corona is given.

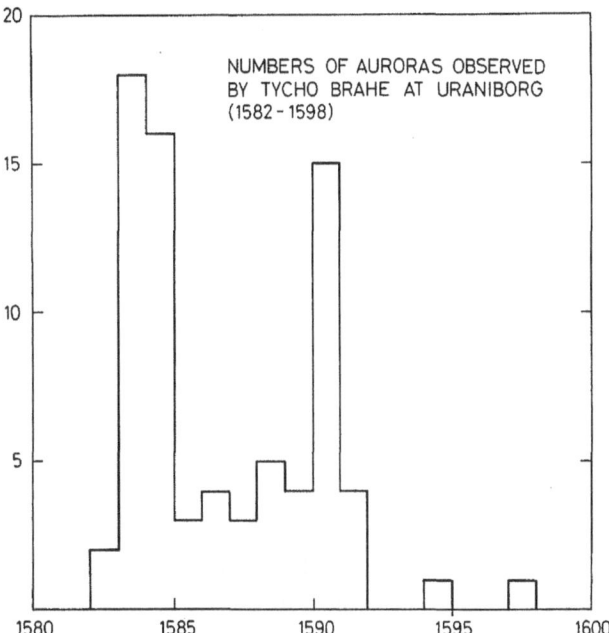

Fig. 4.4. The world-famous Danish astronomer Tycho Brahe (1546–1601) observed several northern lights from his observatory at Uraniborg on the Island of Ven. He was a very skillful observer. Kepler's laws are to a large extent based on Brahe's observations

Fig. 4.5. The annual number of northern lights which Brahe and his assistants observed from Uraniborg in the years 1582–1597

These data were later used to determine the magnetic inclination at Uraniborg. Since very few data are available for determination of the magnetic inclination before the 18th century, Brahe's auroral data must be of great interest to geophysicists. Brahe's data, however, were not very well disseminated and are not mentioned in the catalogues of Tromholt, Rubenson, Fritz, etc. (cf. Fig. 10.3). The annual numbers of the northern lights observed by Brahe are presented in Fig. 4.5. Tycho Brahe did not present any original theory of the northern light and he believed it was due to a sulphurous vapour.

Brahe, however, made another observation of very great importance to auroral physicists in that he proved that the comet he discovered in 1577 was further away from the Earth than the Moon. It became clear to him that comets were not light phenomena associated with the Earth's atmosphere. The word "comet" from Brahe's findings got a new meaning, and from then on was restricted to the same celestial phenomenon as it is today. After Brahe's observation there was no reason for confusion between northern lights and comets in historical data, as there was before 1577. But in spite of that, in a book printed at Uraniborg by Peder Jacobsen Flemløse in 1591,

the words "great comets" are still used instead of "chasmata" or northern lights.

4.7 Absalon Pederssøn Beyer
Made the First Drawings of the Northern Light in Norway

A contemporary of Tycho Brahe in Norway was the bishop of the city of Bergen, Absalon Pederssøn Beyer (1528–1575). In his diary covering the time period 1552–1572 he prepared a card before Christmas in 1563 which was a remarkable drawing that truly depicted the northern light. He was in all likelihood one of the first persons in the world to illustrate the northern light from his own observations. The drawings today seem to be very primitive in their development, and bear a strong resemblance to the mysticism and diffidence with which the northern light was associated in the 16th century.

Pederssøn Beyer was undoubtedly Norway's best-known author in the 16th century. He was born

39

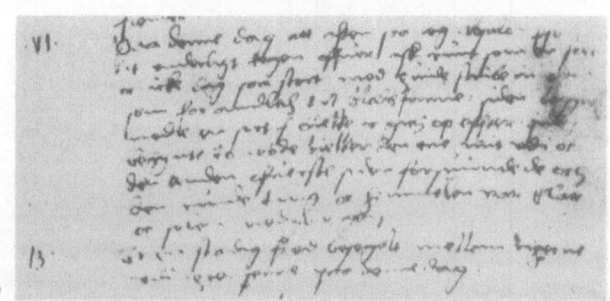

Fig. 4.6. a The first known drawing of a northern light observed in Norway. The illustration was made by Absalon Pederssøn Beyer (1528–1575) sometime in the winter of 1573/74. The northern light is the pencil of rays in the upper left corner where the following words are written: "smoke, fire, vapour". Below this phenomenon are shown dark and white clouds. The illustration clearly shows that Pederssøn Beyer believed that the northern lights occurred above the clouds like meteors and comets. *b* Facsimile of a part of Beyer's manuscript. Concerning the northern light

scriptions in Norway until about the year 1800 were made with few exceptions by priests and bishops.

4.8 Confusing Descriptions of the Northern Light by Peder Claussøn Friis

In his thesis *About Greenland* in 1596 Peder Claussøn Friis (1545–1614) added a passage about the northern light which he had translated from *The King's Mirror* (cf. Chap. 2.2 and Chap. 4.3). The theories presented in *The King's Mirror* were not translated by Friis because he said they were neither right nor exact. There is a great difference in form and content between the description of the northern light in *The King's Mirror* and the one given by Friis, and in particular Friis went further into the dynamic character of the northern light. In addition, the priest from Audnedal, Southern Norway, adds on his own the following account:

"The northern light always appears far to the north and does not reach high enough in the sky that it can be seen at places other than Greenland, Iceland, and in the northern part of Norway. This is why it is called the northern light".

The above has led many to believe that the northern light was not seen from the southern parts of Norway around the year 1600.

Why did Claussøn Friis include this information about the geographic extent of the northern light? An exact answer to this question today is impossible, but one might perhaps understand some of the background by a more detailed study of *The King's Mirror*. The narrator in *The King's Mirror* focusses his total interest on the northern light in Greenland. The theories that are put forth in *The King's Mirror* are tailor-made for Greenland, which is situated near "the utmost borders of the Earth".

Mr. Peder wanted to explain to his readers why *The King's Mirror* did not refer to the northern light in Norway, and also to give a historical reason for the

in Aurland West Norway, graduated from the University of Copenhagen in 1551, and became the palace priest at Bergen House in 1566. He was married in 1553 and after Beyer's death, as a matter of interest, his widow was charged with sorcery and burned at the stake in 1590.

In Beyer's manuscript, the words "in the west, smoke, fire, flame and cusp" are used several times in his description of the northern light. Also where the word "sound" is used auroral sound could be indicated, and in this connection one sentence of special interest is:

"I thought it was the Møllendal's river" (a river close to where he lived).

A second description from Beyer of particular interest is the following:

"January 2, 1564 (the night to Monday in the year which was a leap year) Peder Simonson's servant, who was reliable and a pious person as well, saw the sky open and there seemed to fall from it a glowing Saint Olav's sword. It seemed to fall down towards the earth and then rise again. Shortly afterwards glowing twigs fell from the clouds which looked like a burning bunch of straw."

Six other descriptions are mentioned in which fire falls from the sky, and strange omens and bloodskies appear but they follow approximately the descriptions used in the above examples.

Pederssøn Beyer's contribution to the knowledge of the northern light is limited to these small notations, but it was he who introduced this long tradition among Norwegian theologians – i.e., auroral de-

Fig. 4.7. Peder Claussøn Friis (1545–1614) was a priest in Audnedal close to Mandal in the southernmost part of Norway. He was the first to translate parts of the Edda literature into a Norwegian everyday language. Claussøn Friis is best known for his topographical descriptions of Iceland, the Faroe Islands, Greenland, and Norway

The King's Mirror and where the northern light is indicated as only occurring at the most northerly places.

Unfortunately, Friis' first edition of *About Greenland* has been greatly misinterpreted by many scientists. The last edition could have prevented much of the disagreement about the period and occurrence of the northern light which was so evident during the 17th and 18th centuries.

Friis stated under all circumstances that after 1550 the northern light again began to appear in the southern part of Norway, which is more than 10° south of the present location of the Auroral oval.

4.9 Petter Dass Did Not Mention the Northern Light

The time period from 1645 until 1715 was again characterized by lack of auroral displays. Modern solar research seems to be in agreement that during this 70-year span of time, called the Maunder Minimum, the Sun must have been unusually void of sunspots (see Chap. 9). The climate in Europe was so bad in this period, that it has since been called the Little Ice Age.

A great author and priest named Petter Dass (1647–1707) lived in Norway in this dreary period of time. In spite of the fact that he devoted a whole chapter in his book *Nordland's Trompet* to weather behaviour in Northern Norway, he did not mention the northern light in his writings. It is reasonable to believe that he never observed a real auroral outburst. It is very ironic that by accident this brilliant writer should have lived in such a quiet auroral time and that he should not have been able to experience the inspirations which these mighty, humbling, heavenly displays can create in the human mind.

Despite this scant frequency of northern lights, a Norwegian with the Latin name Christiernus Reitherus in his monograph *Historico Geographica de Orbe Septentrional* or *Historical Geography of the Northerly Hemisphere* written in 1664 discussed the

expression "Norderljos" (Northern light). Notice that Mr. Peder mentioned both Iceland and Northern Norway in his translation of *The King's Mirror*, neither of which occur in the original text. We have not been able to find out if Claussøn Friis had original sources not known to us, or if he only assumed that the northern light could be seen from Iceland and Northern Norway.

In a revised edition of the treatise *About Greenland* which was written in 1604 or 1605, he added the following footnote:

"This northern light which is now mentioned, has not in ancient times been seen in places other than the Nordic countries, which has been stated earlier. But in the time of my childhood around 1550 it was first seen by those who lived in the southern part of Norway, not higher, however, in the sky than the "Nordledingen". [13]

But now since anno 1570 it (the northern light) comes so high in the sky that it is seen southeast and south of us, and I think that it now also might be seen in other countries".

The footnote therefore gives almost the directly opposite impression of that contained in the famous sentence which Friis added to his first translation of

---

[13] Nordledingen is the polar star and the leading star, the same star that shows the road towards the north

Fig. 4.8. Peter Dass (1647–1707) was a priest in Alstadhaug in Northern Norway. He was also a skillful writer and his best-known book is *Nordland's Trompet* in which he describes the climate and way of living in Northern Norway. In his description of the climate he surprisingly enough does not mention the northern light, even though it is known that the fishermen in the area where Dass lived used the northern light as a weather sign. Since Dass lived in a period with probably very few occurrences of the northern lights [the Maunder Minimum (1645–1715)], it is possible that he never saw this phenomenon

phenomenon in a little section of his monograph. He quotes Claussøn Friis and makes use of the designation Lumine Septentrionali, which directly translates as the words northern light. Reitherus had often observed the northern light and speculated on its cause. He was convinced that it arose because of some kind of light refraction, but he could not shed any further light on the cause.

Reitherus' assertion that he had often seen northern lights is not necessarily in conflict with the fact that there were very few northern lights to be seen in the Little Ice Age. Reitherus says nothing about when he saw the northern lights and the Little Ice Age had still not fully begun in 1664 when the monograph was written.

Another interesting piece of information about the northern lights in Norway in the 17th century, which adds to the rather scarce written sources of that time, is an account of a journey written by the travelling Italian merchant Francesco Negri who sailed to Norway in the years 1664–1665 (cf. Chap. 3.5). He wrote a very detailed description of how the northern light behaved and was highly impressed by what he experienced. He also associated the northern lights with the very rare phenomena seen at the temperate zone which he called "capre saltanti" (jumping goats) or "tizzoni ardenti" (flaming trees). He did not believe that those were the same, since the northern light, according to Negri's belief, was caused by

an emanation from the ground. In order to contribute to the search for an explanation of this emanation he mentioned a special kind of peat growing in the mountains of Norway and which is very rich in hydrocarbon liquid. Negri commented that this hydrocarbon vapourizes in the summer but is prevented from doing so in the winter because of the strong cold. In the winter therefore the hydrocarbon vapours will penetrate the soil downward all the way to the bedrock where they will be ignited and fly out again as long fire strips. Negri is, however, not totally firm in this idea as he writes at the end of his discussion,

"but I would very much like to learn about someone better informed, if they could give me a more satisfactory explanation about this phenomenon".

4.10 Norway's First Historiographer, Thormod Torfaeus Discusses the Northern Light

In 1706 Thormod Torfaeus (1636–1719) wrote a book having the title *Grønlandia Antiqua* or *Ancient Greenland*. In this book he mentions the northern light and gives descriptions of it in the Old Norwegian language. In the following English translation, we place more emphasis on the substance than on the syntax:

"It seems like a very big, wide flame that instantaneously had been ignited in the northern parts of the sky, and stretched out like a high and long mountain, the way that the wind in one pipe of an organ goes out for a moment, from another pipe to the next, up and down, it rises in height for a moment, and then instantaneously goes down low. Those who have not seen this cannot believe how fast it moves about. When the moon changes phases a new northern light will be ignited. During the darkest nights, it shines and glitters as brightly as it does outdoors and through the windows of houses on a clear full moonlit night. But when the Moon is full, the northern light does not appear so bright when it, in its greatest brilliance, it appears as if it jumps and leaps around the sky in the north, then suddenly it is extinguished, darkens and looks like smoke and cloud, and then again it brightens up in other places where it has just seemed to have been extinguished, it ignites and flies around with the same brilliance and speed as before. At the break of day it disappears".

No doubt Torfaeus must have observed many northern lights, for only a person who has seen the magnificence and the grandeur of these heavenly lights could describe them so clearly.

Torfaeus took *The King's Mirror* to task and pointed out that the northern light is not a phenomenon which is confined to Greenland only since it also occurs in Iceland, the outermost parts of Norway, and also very often at Karmsund which is situated at 59° north geographic latitude where he lived for the last part of his life. He suggested, nevertheless, that since the northern light appears to be ignited at greater heights over Greenland than in other places, it is possible that it had first begun in the northerly areas and

Fig. 4.9   *Fig. 4.10* ▶

Fig. 4.9. Thormod Torfaeus (1636–1719) a historian from Iceland, was employed by the Danish King to translate Norse literature to Latin. He became a tax collector in Stavanger, Norway in 1664 and lived on the island of Karmøy the rest of his life. Torfaeus collected several of the Sagas of the Icelanders and brought them to Denmark. In 1682 he became a well-paid historiographer in Norway, but his historical work was never considered to be very reliable

Fig. 4.10. Ole Christensen Rømer (1640–1710) a Danish astronomer. His observations on Jupiter's moons enabled him in 1675 to determine the speed of light. He made many improvements to observational techniques in astronomy. He was also in charge of introducing the Gregorian calendar in Denmark and Norway in 1700

Fig. 4.11. In 1707 the northern light was seen from Copenhagen on February 1st and March 1st and 6th. The Danish astronomer Ole Rømer observed these lights and made the illustrations shown below Rømer's illustrations are probably more realistic than many of the contemporary drawings of the northern lights. The horn-like structure is very reminiscent of the "jumping goats" mentioned by Aristotle. (The figures are taken from the thesis by J. F. Ramus, see Chap. 5)

43

then broadened more toward the south. Following along the same line of thinking, he pointed out that Claussøn Friis must have been mistaken when he claimed that the northern light was not high enough in the sky to be seen farther south than Greenland, Iceland, and the most northerly parts of Norway (Chap. 4.6). Regarding his own experiences as a child in Iceland, he recalled that the northern light often put such fear in people that they "fell under a spell".

It is difficult to determine what Torfaeus meant by saying that "the northern light often could be seen from Karmsund". It is doubtful that the northern light moved to greater southerly latitudes any more often during the year when Torfaeus lived at Karmøy than in any other. At any rate, during the night between 1st and 2nd January 1707 it is known that the northern light was seen all the way south to Copenhagen and viewed there by the Danish astronomer Ole Rømer (1640–1710). In his description of the event Rømer began with two illustrations. He drew an auroral arc that had some crescent-shaped horns which simply reminds one of Aristotle's usage of the descriptive words: "goats" and "jumping goats". Regarding the occurrence of the northern light Rømer says that "it is seen in Norway and Iceland almost every single year".

All pictures on p. 45 to 51 are made by Steinar Berger, The Auroral Observatory, Tromsø

Right Folding rayed band;
Below Sunlit band. Weak rayed structure

Right Partly parallel bands;
Below Rayed band with red lower border (Type B)

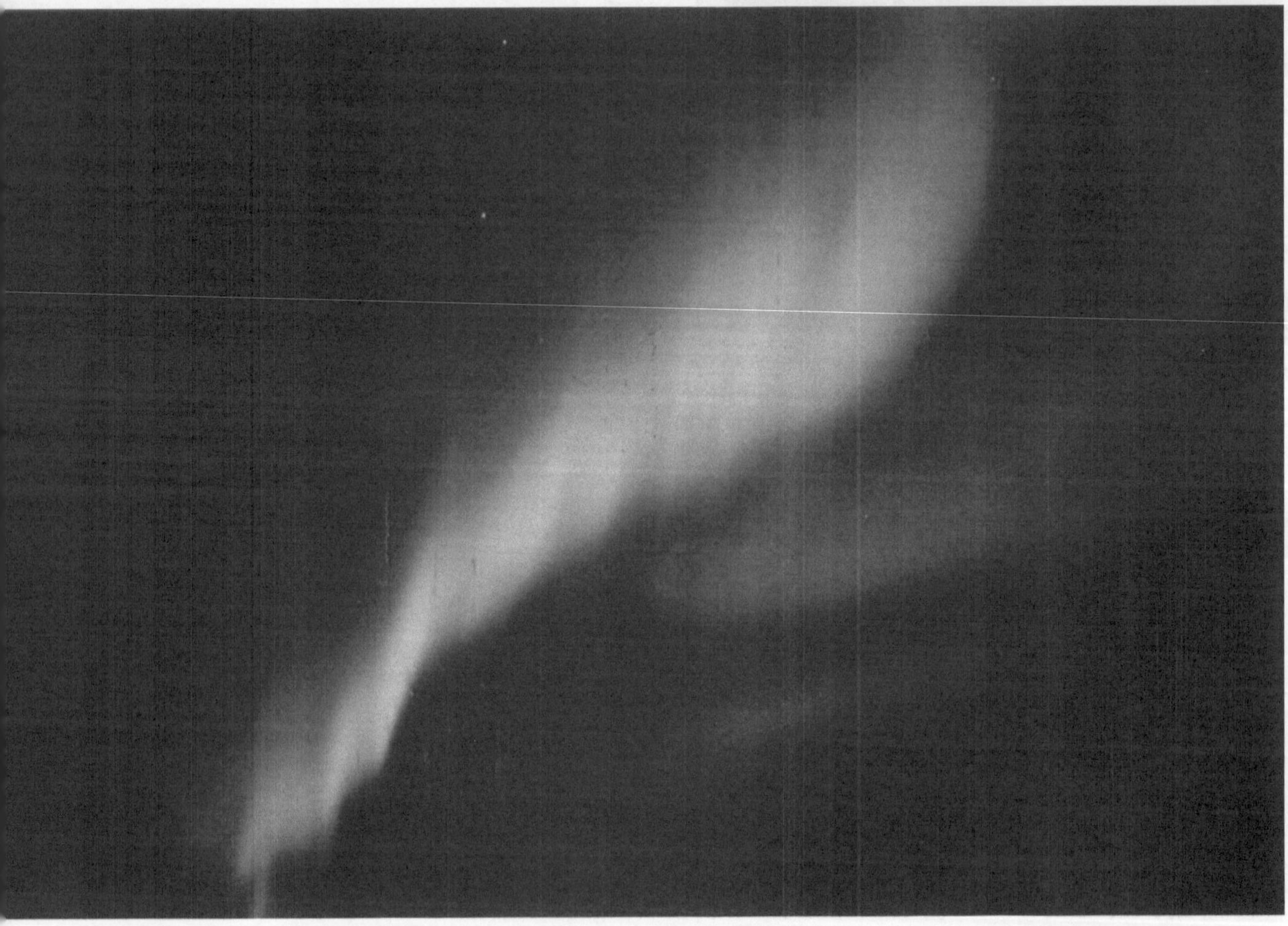

Left Auroral rays; *Bottom left* Parallel homogeneous arcs, the upper one is a rayed arc

Below Homogeneous arc; *Bottom right* Active rayed band

Next page
Upper part Homogeneous arc partly rayed to the left; *Bottom* Rayed auroral band (drapery)

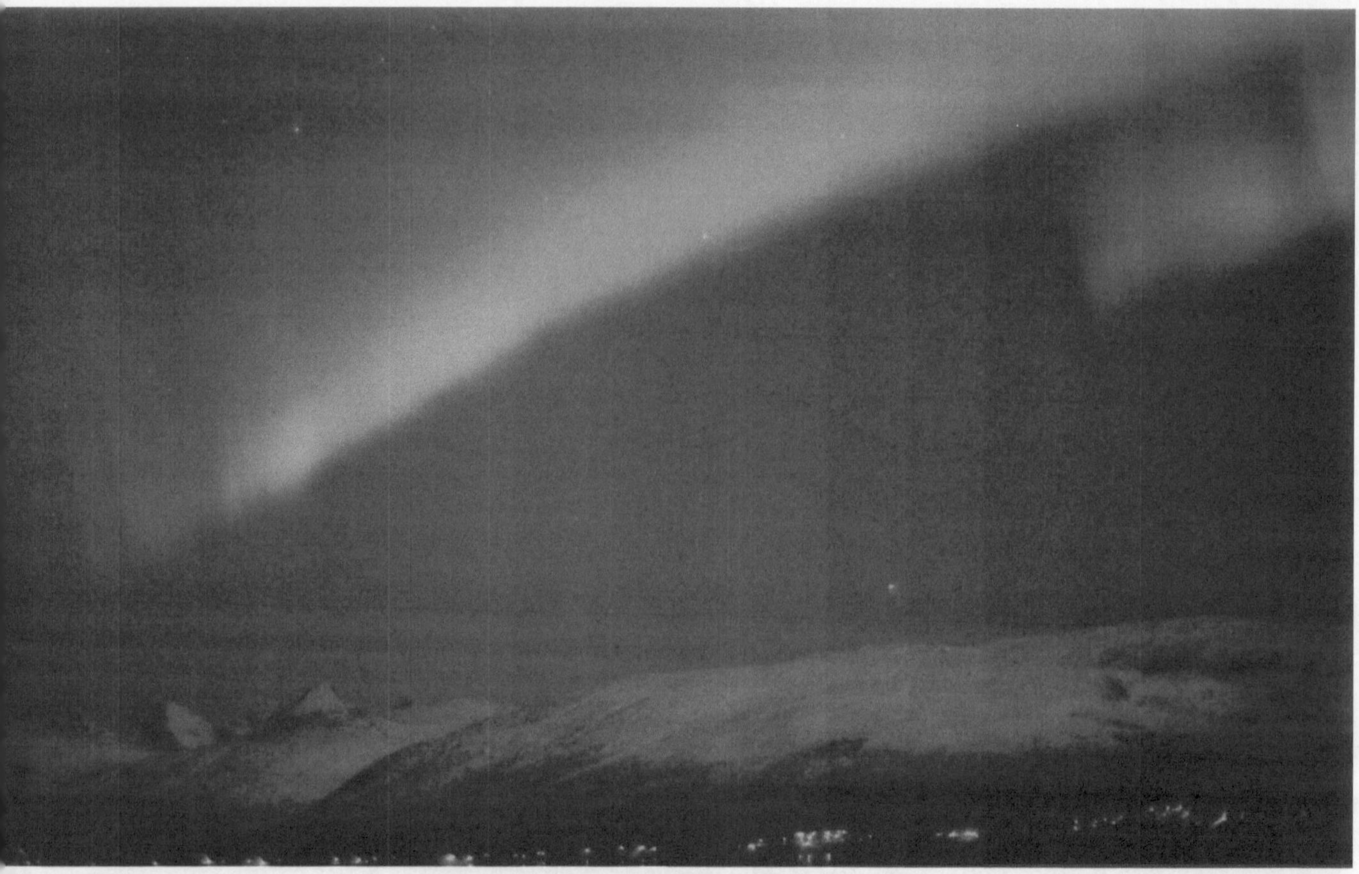

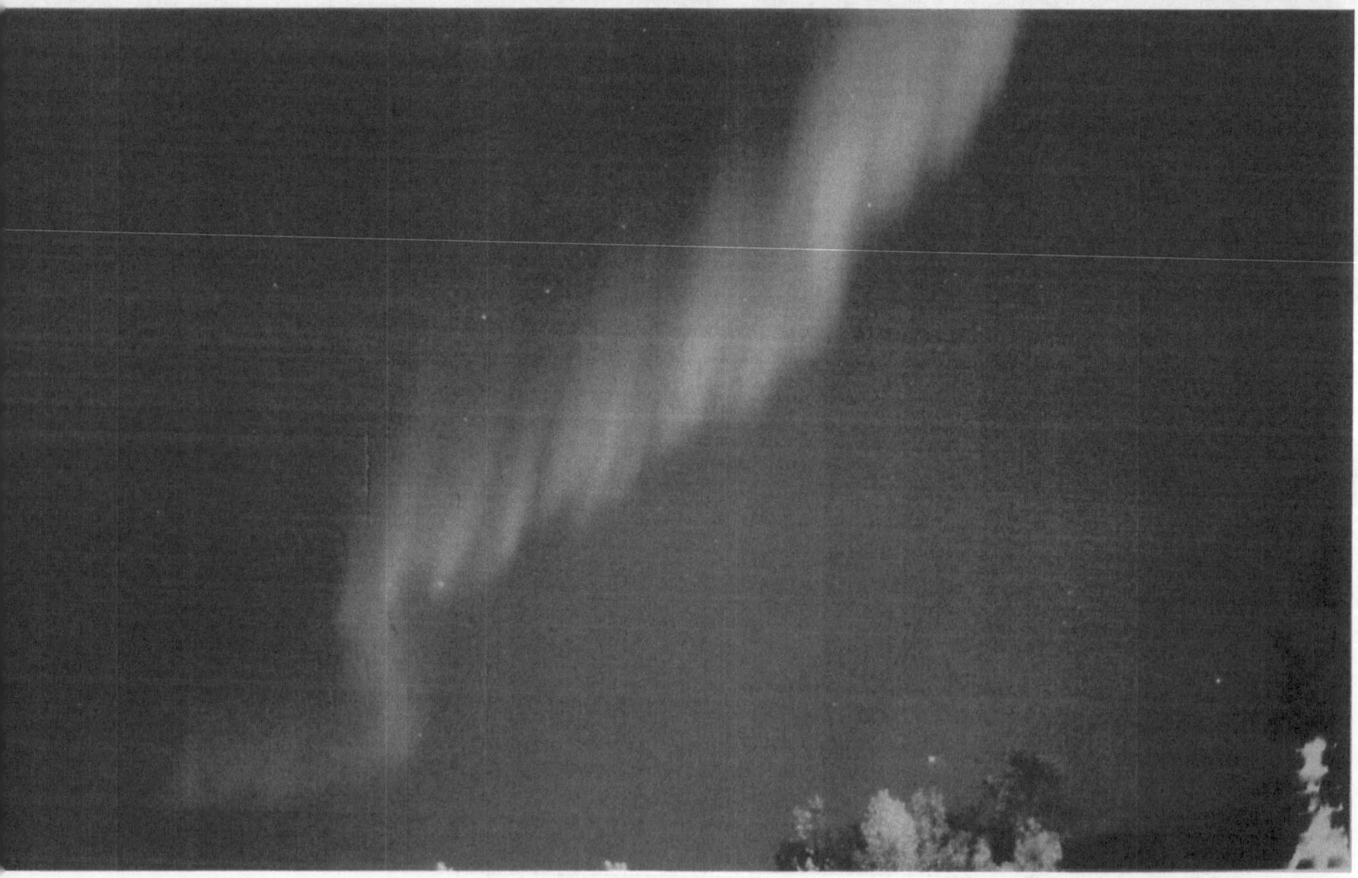

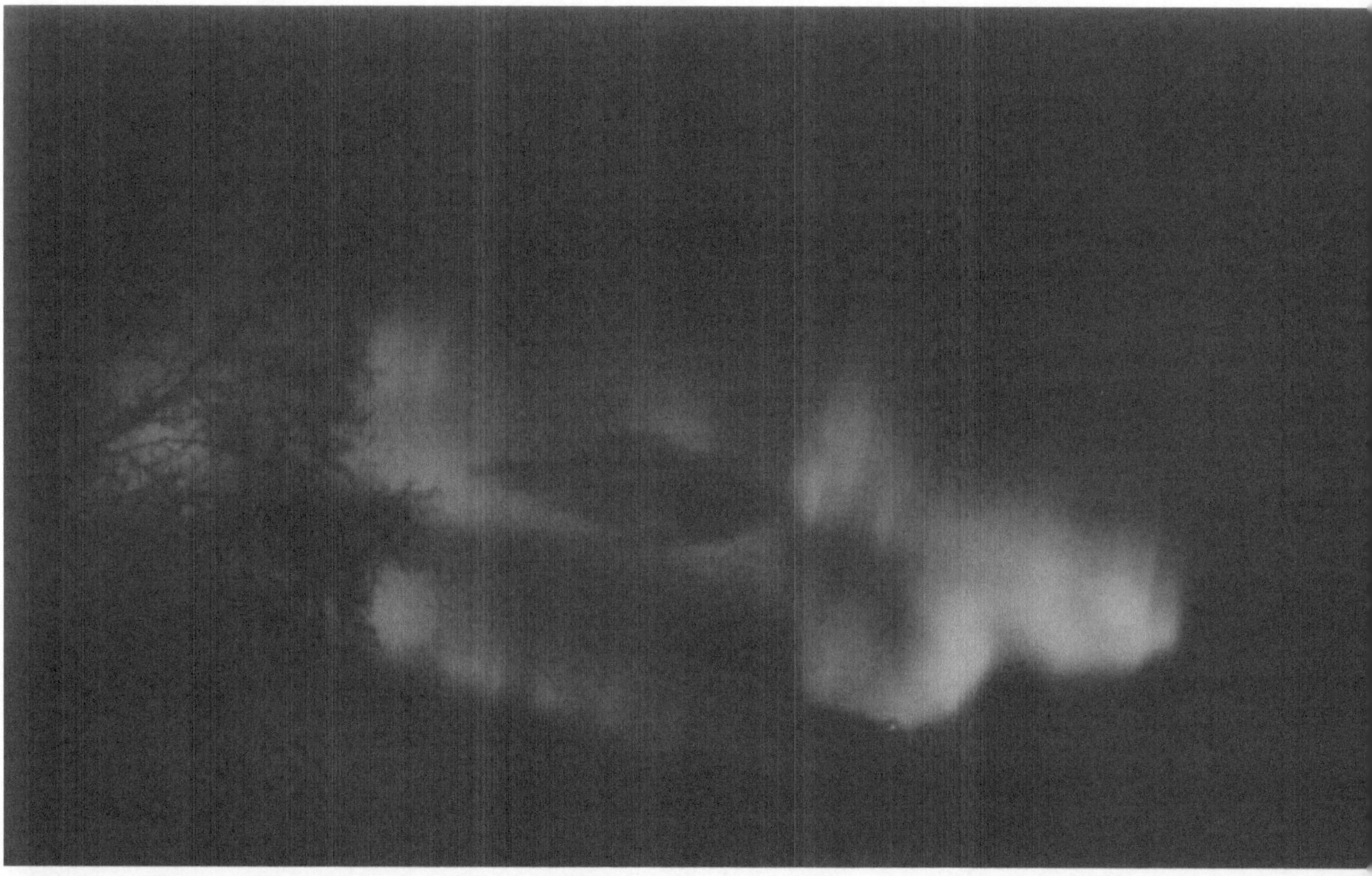

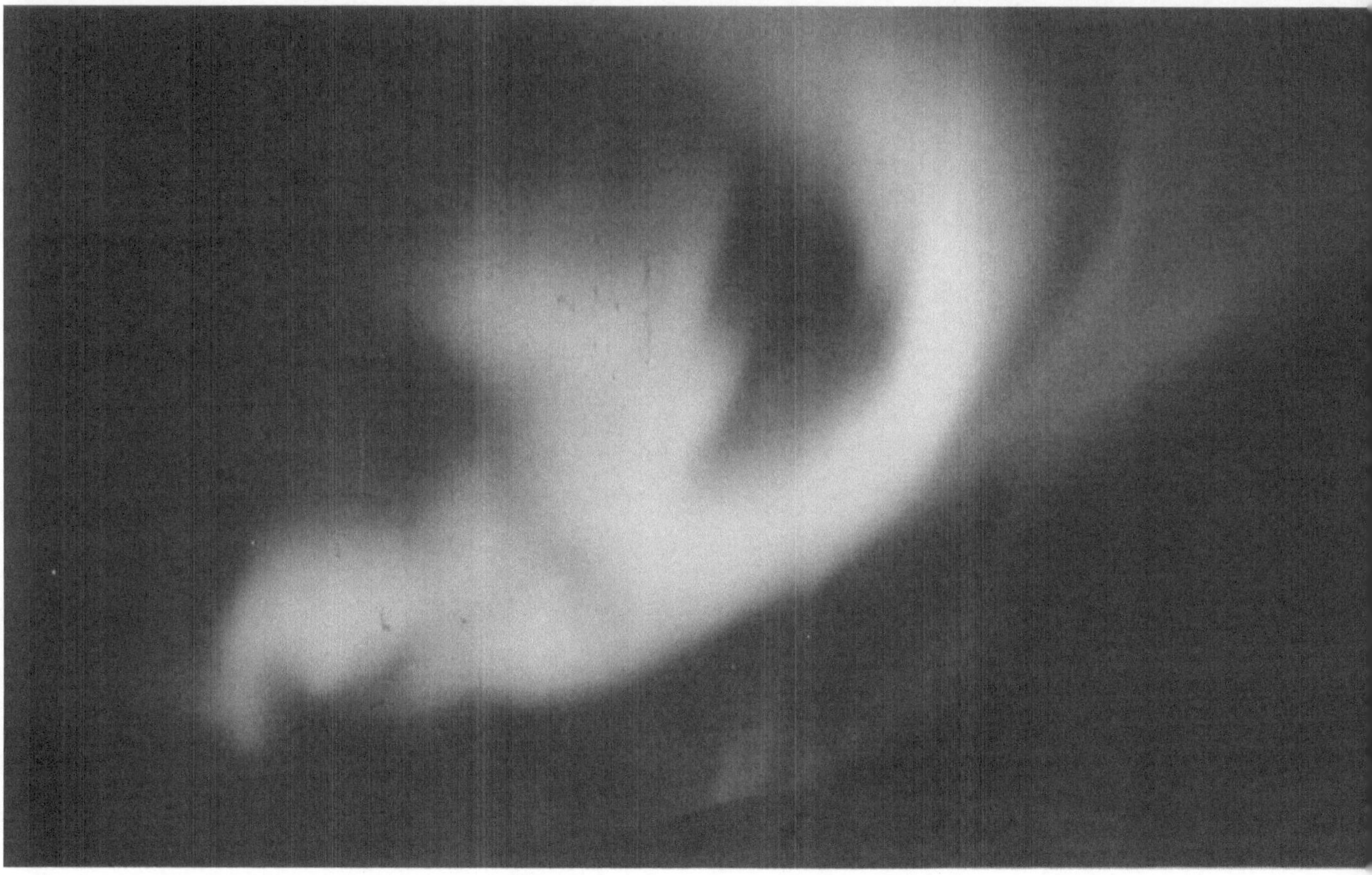

Preceeding page
Upper part Active rayed bands; *Bottom* Active looped band with
a strongly lower red boarder (Type B)

Below
Several rayed bands during break-up. Note the intense loop
structure in the *lower left corner*

Next page
Very active auroral displays with complicated structure. Note
the illuminated landscape in the *bottom panel*

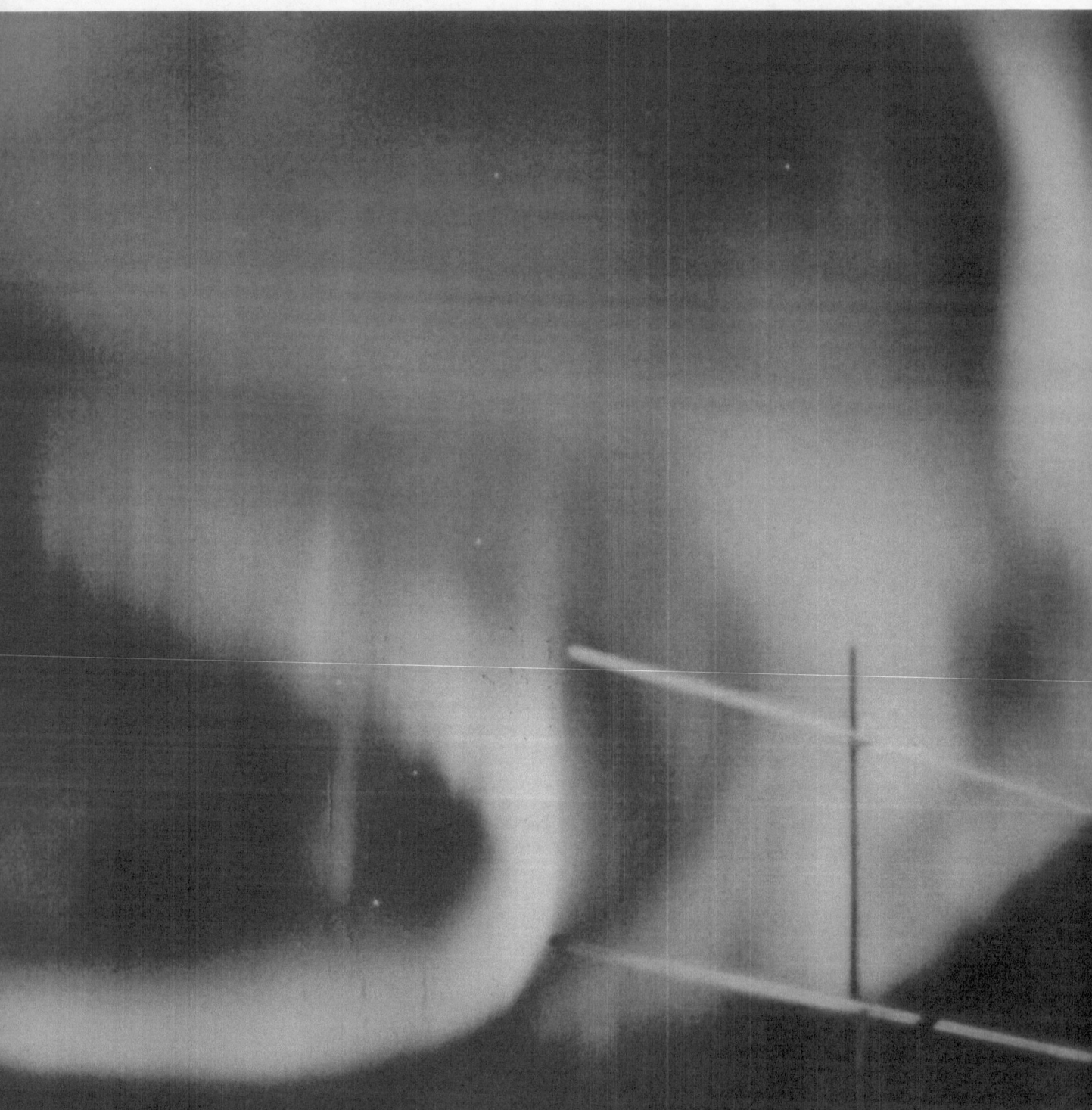

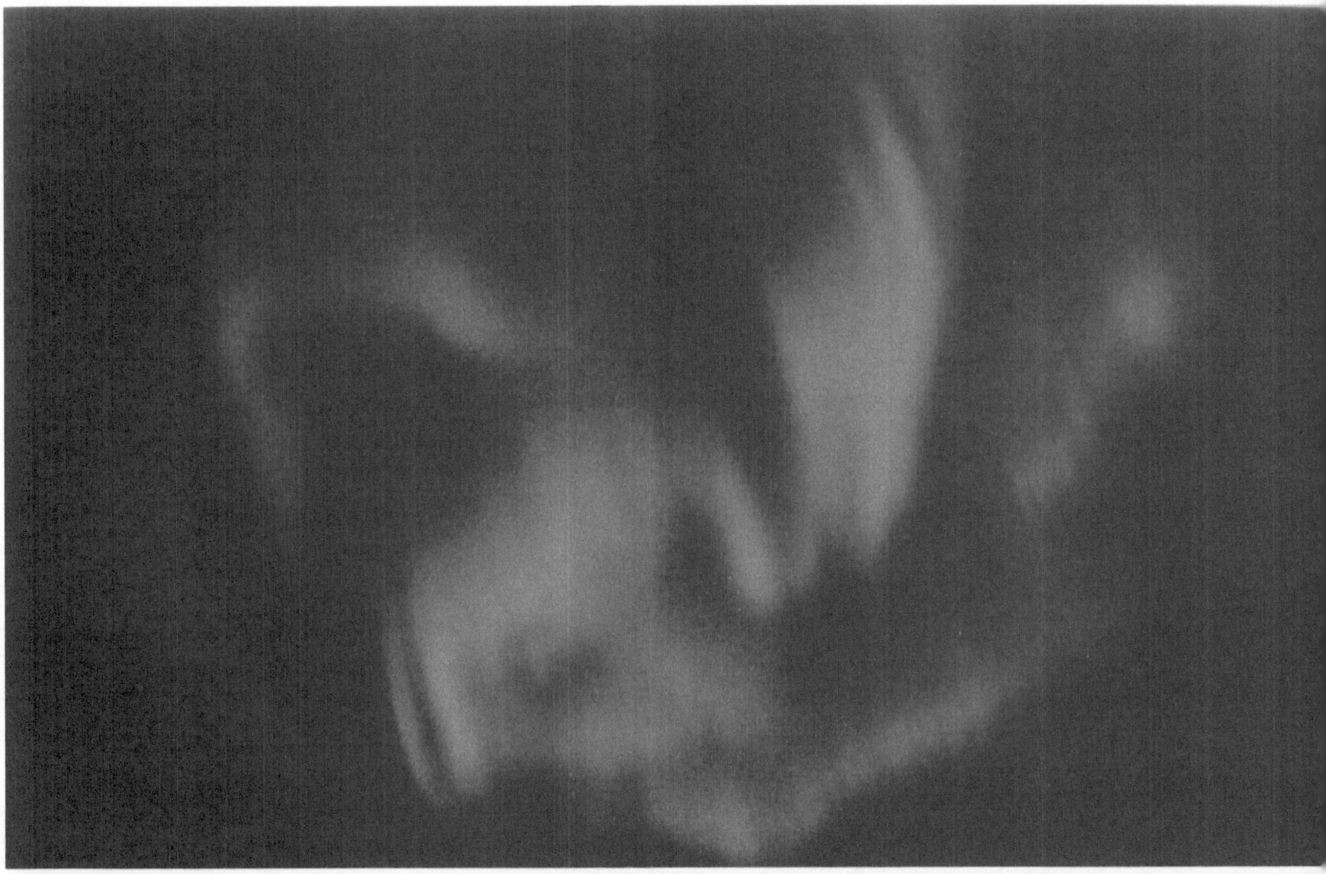

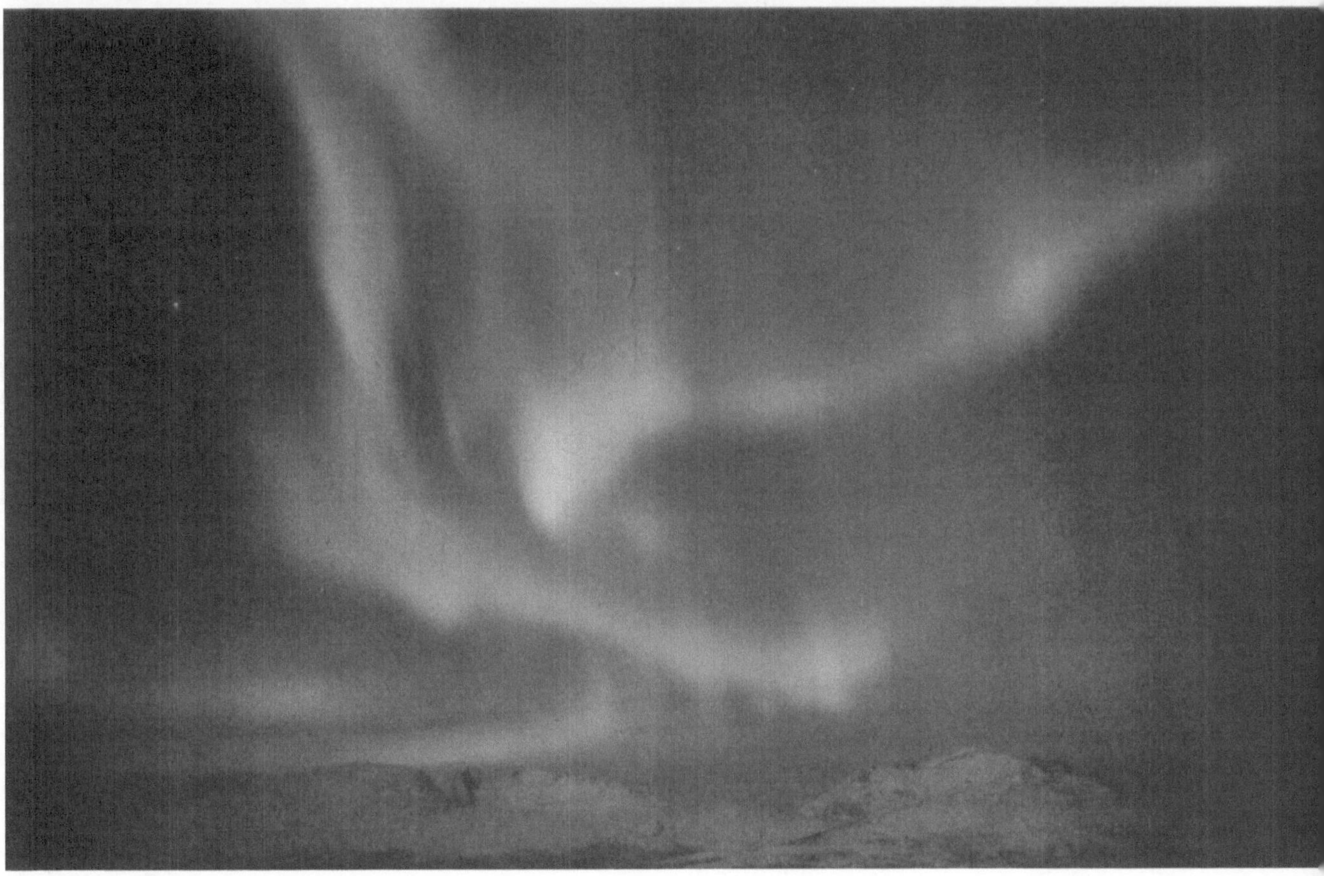

Paintings of the northern light by well-known Norwegian artists. *Upper panel* Peder Balke (1804–1887). Rayed auroral form painted about 1850

Bottom panel Aquarell by Gerhard Munthe (1849–1929) from 1892. The picture is entitled "Suitors" or "The Auroral Daughters" which shows that the artist probably was aware of the old tradition relating the northern light with dead maidens (cf. Chap.1)

5 The Northern Light in Scandinavia During the Eighteenth Century

5.1 Suno Arnelius Wrote the First Scientific Treatise on Northern Light in Scandinavia

In 1708 a Swedish priest, Suno Arnelius (1681–1740) submitted to the University of Uppsala a manuscript bearing the Latin title *Exercitium Philosophicum de Chasmatibus* – a philosophical treatment of the northern light. Arnelius, however, never got the chance to defend his thesis.

He contended that the northern light is caused by vapour or gases of a special kind which rise from the Earth. In the polar regions where the air is cold, ice particles are formed out of these gases. When the Sun is below the horizon, rays of light strike the ice particles, and these rays are reflected in such a way that they can be seen in the sky from some distance away.

Arnelius was convinced that ice particles must play a significant role in producing a northern light since it seemed to him that a special type of snow crystal appeared to be falling under the northern light. And since northern lights are seen only during the winter-time, he thought this could result from the fact that ice particles melt during the summer. His conclusion therefore was that the northern light must be a reflection phenomenon.

He used two drawings (one of them shown in Fig. 5.1) to illustrate why a number of small irregularly distributed ice foils should give a diffuse rather than a sharp picture of the Sun as would a flat plane mirror in the atmosphere. On this basis, the structure of the northern light would be determined by the manner in which the ice particles are distributed in the atmosphere.

He also maintained that since the northern light is seen only in clear weather its height must be greater than the clouds. Winds played an important role in blowing the ice crystals away causing the northern light to disappear. In this connection Arnelius wrote: "Eliminating the source, eliminates the effect".

Ice particles are confined to specific layers in the atmosphere, and when these layers move relative to each other, movement of the northern light will occur, Arnelius claimed. The smallest waves in a layer will give a double moving effect in the northern light in the same manner as a reflected picture from a moving mirror appears to move with twice the speed of the mirror.

He maintains that the northern light can move either from east or west towards the north. Because the Sun's rays must pass through the atmospheric layer under the ice crystals twice, the rays will be refracted into colours. Any clouds that happen to be in that layer of the atmosphere will also be affected by the light refraction. In this manner, Arnelius explains the colours seen in the northern light. "As anyone sees his rainbow, so will anyone see his northern light", Arnelius wrote.

If ice crystals form in the zenith, they will cast light rays in all directions such that the northern light appears to radiate out from the zenith. Such zenith

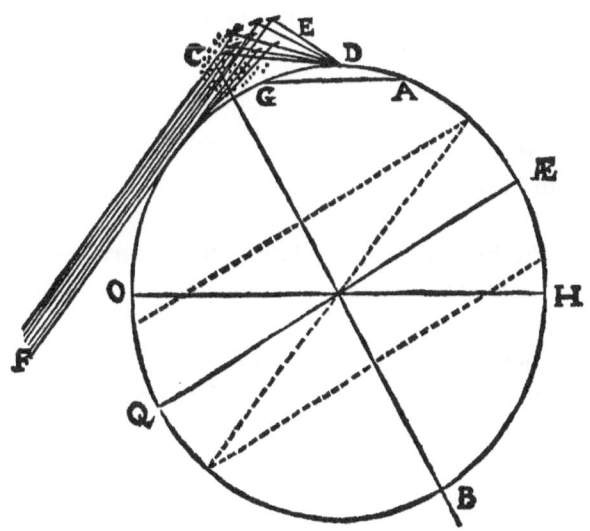

Fig. 5.1. Suno Arnelius made this illustration in his thesis to explain the northern light. *CF* indicates the solar rays which are reflected by the ice particles in the air, and the reflected light can be observed at the Earth's surface

Fig. 5.2. Jonas Ramus (1649–1718), priest in Norderhov, Norway from 1690 and married to the heroic Anna Colbjørnsdatter

aurora, however, do not belong to the class "Chasmata" and they seldom occur there because the basic materials causing the northern light are practically never in the zenith.

In his conclusion Arnelius exclaims, "I have explained the northern light", and believes that no new observations can cast new light about the cause of the northern light. He cautions against following the multitudes blindly in viewing all unusual events as miracles and prophesies about bad things to come – the same problem which Olaus Magnus had faced and cautioned against 150 years earlier.

Without taking anything away from Arnelius' originality, it must be remarked that his theory reminds one very much of the second hypothesis given in *The King's Mirror* (Chap. 2.2) rephrased to fit in with his up-to-date refraction theory as applied to the northern light.

5.2 Jonas Ramus Associated the Northern Light with Magnetism

In the beginning of the 1700's, the occurrence of northern light became so commonplace in Scandinavia that people began to question the idea that the northern light predicted dire consequences and turned instead to search for its real significance. Into this state of affairs, there came the priest from Norderhov, Jonas Ramus (1649–1718) whose wife Anna Colbjørnsdaughter was the heroine in the battle of

1761 between Sweden and Norway. In 1715 he published a book *Norriges Descriptions* in which one chapter was devoted to the northern light. He begins by mentioning the northern light of 1707, which was seen over Copenhagen, and describes it as looking like many organ pipes standing beside each other from which air was blown upwards in many blasts after which the air quickly fell down again.

Ramus is steadfast in the idea that the northern light cannot be reflections of sunlight or moonlight, and therefore disagrees with that particular hypothesis in *The King's Mirror*. He dwells on the theory held by some people that there should be subterranean heat in Greenland which would make grass and plants grow rapidly and prevent water from freezing during the winter. Assuming this heat is located inside the Earth, there must be a hole or opening in the mountain where the smoke is able to come out into the air. As an indication that such a hole can actually be found, Ramus mentions the fact that "the compass" or the magnet always points towards the north. This would mean that a mountain of iron should be located under the north pole.

The Earth is frozen and snow-covered because of this very cold environment. Under such conditions, the iron material in the Earth cannot evaporate any more. Ramus feels that the hole from which magnetic material comes and from which "the compass" gets its sustenance must exist.

Ramus' speculations contain original ideas about magnetic strengths. He expresses doubt that the northern light is the result of burning, spewing fire, but also notes that the rapid motion of the northern light is difficult to explain by his theory.

At the present time, such ideas may appear to be a lot of nonsense, but one must judge them by keeping in mind that contemporary researchers were groping for an understanding of the northern light. At this time Ramus had enough doubts about one of the theories in *The King's Mirror* and the courage to reject it as unacceptable, in spite of the fact that no better hypothesis had been put forth in 500 years. Moreover, he had the originality to give new momentum to the theories contained in *The King's Mirror*. Being a priest in the countryside of Norway, isolated from the rest of the world, and in particular from the scientific centres in London and Paris, it is a most interesting fact that Ramus had discussed the Earth's magnetism only about 100 years after Gilbert in London had discovered that "The Earth is itself a gigantic magnet".

Ramus did not have enough insight into physical processes to see the very close correlation between magnetism and the northern light; that would be left instead for the famous Swedish astronomer Anders Celsius and his assistant Olof Peter Hiorter (see Chap. 5.8 and 5.9).

Only a year after Ramus had completed his book, a very large northern light occurred in London on 6th March 1716 which the famous English astronomer Edmund Halley (1656–1742) observed. In *Philosophical Transactions* for March 1716, Halley gave a detailed description of the phenomenon based on his own and other observations. He expressed an almost romantic enthusiasm about having personally observed the event and remarked that after 60 years he had almost given up hope to ever get the chance to see this phenomenon; the only one of the classical light phenomena of the sky which up until then he had not observed.

Just as Ramus had done previously, Halley sought to explain the northern light by assuming that underground heat evaporates water and that the resulting steam rises from the gases into the atmosphere. Like an earthquake, the northern light occurs on an irregular time schedule and it presented to Halley a problem more capricious than any other natural phenomenon with which he had previously dealt. He found it natural to seek a common source for these two phenomena. In Halley's time it was generally believed that sulphurous vapors rising from the Earth's interior meltingpot was the cause of earthquakes.

The 6th March 1716 aurora seen in London was also observed all over northern Europe, from Ireland in the west to the border between Russia and Poland in the east.

When no earthquake was reported at this time from any place in northern Europe, Halley like Ramus rejected the "heat theory".

Rather, Halley hypothesized that the northern light is due to magnetic evaporation from the Earth's interior. By describing an experiment in which one places a magnetic sphere on a plate sprinkled with iron filings, Halley illustrated how magnetic material aligns itself around a magnet and permeates this to a somewhat greater degree than at the magnetic equator. He believed that this volatile matter which now and then streams out of the magnet could pro-

duce a northern light just as electric matter can produce light.

From knowledge of the deviation of the magnetic needle, which in London around 1715 was 12° west for this geographic meridian, Halley concluded that Norway lies further from the magnetic north pole than both Iceland and Greenland. He thought that the northern light was most concentrated around the magnetic pole, and therefore would be seen less frequently in southern Norway than for example in Greenland.

If one compares the speculations of Halley and Ramus it is easier to agree with those of Halley. But one must remember that Ramus did not have the advantage of living in London which was one of the world's centres in sciences, and where such geniuses in physics and natural science lived as Isaac Newton (1642–1727) and Robert Hooke (1623–1703).

Ramus' ideas are, however, quite original and this originality illustrates how well the local priests in Norway were up to date on current problems.

5.3 Jens Christian Spidberg, the First Scandinavian Who Wrote a Book on the Northern Light

The first publication which was entirely devoted to the northern light and written by a Scandinavian was printed in Germany in 1724. It bore the particular title *Historische Demonstration und Anmerkung über die Eigenschaften und Ursachen des sogenannten Nord-Lichts* and was written by Kristiansand's bishop Jens Christian Spidberg (1684–1762), a prominent naturalist and cartographer.

Spidberg's real incentive for writing the book was Descartes' (1596–1650) work *Meteora* from the first half of the 1600's.

In spite of the fact that Descartes never observed a northern light he put forth a proposal to explain the phenomenon. Descartes proposed that, to begin with, there is an abundance of wind clouds which gather in the air in layers, and air particles are exposed to strong attraction and excitation such that lights are formed in about the same manner as light is produced in a thunderstorm. Descartes' idea was also that the northern light is produced by a strong terrestrial flame which is reflected from the clouds. Spidberg rejected both of these ideas, mainly because the northern light is seldom visible when there are clouds in the sky, and in addition because a large amount of heat would dissipate the clouds to such

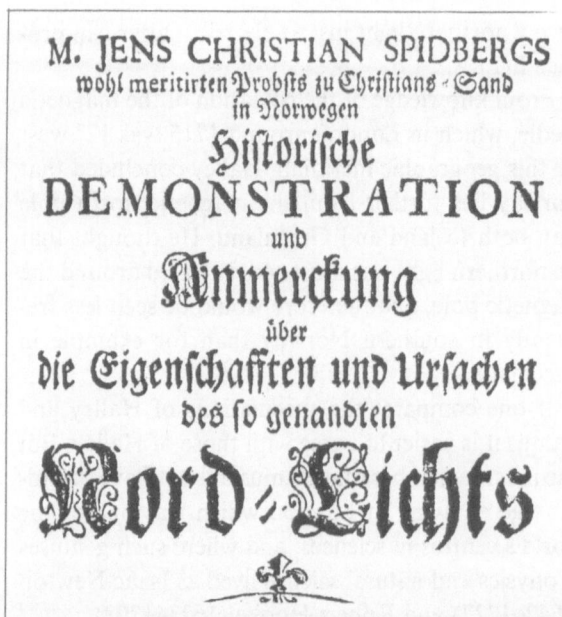

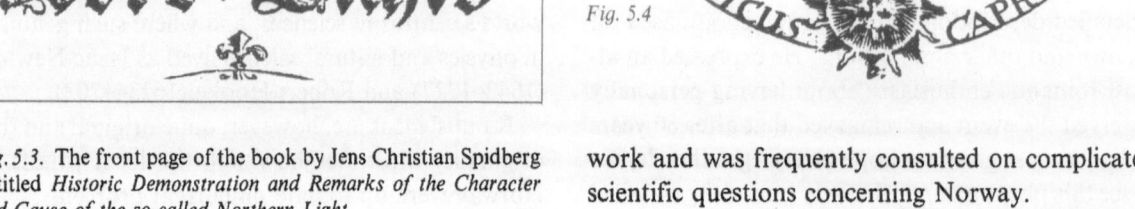

Fig. 5.4

Fig. 5.3. The front page of the book by Jens Christian Spidberg entitled *Historic Demonstration and Remarks of the Character and Cause of the so-called Northern Light*

Fig. 5.4. Spidberg, like Arnelius, presented a theory of the northern light as being caused by reflections of the solar rays from ice particles in the air. This figure, which was originally published in his book, can be compared with Fig. 5.1. The solar rays are reflected in the atmosphere around the Earth marked with circle *A*, and the Aurora Borealis can be seen on the nightside of the Earth

an extent that light reflections would not be produced.

Spidberg sets forth a second theory which is not essentially different from Arnelius' hypothesis (Chap. 5.1), namely that the northern light is due to reflections of solar rays from ice particles in the air. With a detailed figure (see Fig. 5.4) he tries to explain geometrically how the effect is thought to be produced and which angles of refraction actually occur.

Spidberg was in charge of one of the largest private libraries in Norway in his time (ca. 6,000 volumes), but it appears that he used very little time to study any ideas about the northern light other than those of Descartes. He certainly had an opportunity to read Ramus' repudiation of the reflection hypothesis (published 9 years earlier), but Spidberg used his arguments to support the reflection idea. This treatise on the northern light gave Spidberg a high reputation. He was the first Norwegian priest who was awarded the Doctor of Theology degree by the University of Copenhagen, and also the first Norwegian to be chosen as a member of the Scientific Academy in Copenhagen. He was very dedicated to his

work and was frequently consulted on complicated scientific questions concerning Norway.

A book on the northern light was published in Paris in 1733 which would later inspire both laymen and scholars to become involved in discussions about the aurora's many puzzling features. It was written by the Frenchman de Mairan (1678–1771) and entitled *Traité Physique et Historique de l'Aurore Boreale* which was inspired by a northern light which he had observed in Paris on 19th October 1726. In this book he rejected the theory that the northern light is caused by zodiacal light – the faint rays of light which stretch out from the sun along the ecliptic.

The zodiacal light is yellowish in color and is sunlight which is reflected from a cloud of solid particles that are concentrated between the planets in the ecliptic plane. In north Scandinavia it can only be observed after sunset, close to the time of the vernal equinox and before sunrise near the autumnal equinox.

The new idea held by de Mairan that the northern light is the result of interaction between the Sun and the Earth's atmosphere represented an important step in the right direction for a better understanding of the phenomenon. His studies of the northern light were based more on what we today would call good scientific disciplines rather than speculation, but he still took a strong stand for a mean height of the northern light at about 800 km.

Spidberg, however, remained sceptical about de Mairan's theory and in a letter to Bishop Erich Pontoppidan (Chap. 5.12) in Bergen in 1750 he wrote the

Fig. 5.5. The impressive title page of the book by Johan Heitman, *Physical Considerations of the Sun's Heat, the Air's sharp Coldness, and the Northern Light*

following: "If Monsr. de Mairan had had some accurate observations on the northern light from Norway one would have expected that his beautiful *Traité Physique de l'Aurore Boreale* would have been more accurate and decisive; because Norway, and in particular the Trondheim diocese, is the ancestral home of the northern light".

5.4 Johan Heitman – A Layman's View of the Northern Light

In 1741 a very interesting book was written by a sea captain named Johan Heitman (1664–1749). The book had a lengthy and descriptive title *Physical Considerations of the Sun's Heat, the Air's sharp Coldness, and the Northern Light*. Heitman was born in Rana and got his mate's training for the sea in Bergen. He was commissioned by the vice-regent to make maps of the coastline of Norway. Heitman had a wide range of interests which included preparing a new calendar and a hymnbook for seamen which turned out to be a best-seller.

The interesting book mentioned above, which to a large extent was based on his own experiences, sets forth a theory that the northern light is caused by nitrous gases in the highest air layers (i.e., to Heitman) between 10 and 20 km. These gases were set in motion by the strong cold winds in the polar region. In much the same manner as phosphorescence is created in salt water by the passage of a ship, so is the northern light created due to movement of the air. Variations in the density of the nitrous gases cause the northern light to appear only periodically. Heitman thought that in the cold of winter the northern light would occur less frequently because when the air is heavier it descends to lower heights below the sulphurous gas layers. He thought that the northern light was not observed earlier in time (during the Little Ice Age) because of the cold weather (Fig. 2.9 and Chap. 8).

In his book Heitman criticized Newton's hypothesis on the velocity of light and the Sun's heat. He could not agree with the idea that the heat which streams out from the sun should be stronger the nearer the Earth is to the Sun. On the whole, he dissociated himself from his contemporary scholars in

physics and mathematics, who he characterized in the following way:

"It would be better, if one abandoned such useless fantasy which gives only the wise ones a cause to be addicted to even larger delusions, as one unfortunately often experiences. In spite of the fact that such scholars report and make examples of such unbelievable matter, I conclude with David: People is nevertheless nothing, great people also go astray, and they count less than nothing, as many as they are".

Heitman died before the book was published and it was his son with the same name who saw that this was done. Possibly because of an expectancy that reactions to the book would be harsh and unkind, the son had a friend to write a poem about his respected father. The poem which was placed in the back of the book closes as follows:

Therefore believe me my good Friend.
If the Book is read by Thee
who unbiased Judge will be,
then he will judge like this (as he rightful can).
Your departed Father was a clever and intelligent Man.

Heitman's work was never received with popularity in educated society, and the translation which Holberg suggested, was so far as we know never done.

Finally, Holberg ends his epistle with the following recommendation about Heitman's writings:

"My naive assesment of this writing, in which I find a new system, is that even if it is not probable, it should not be ridiculed".

It is clear that Heitman's courage and confidence, being great enough to criticise geniuses like Newton and Descartes, must have created a certain admiration in Holberg's house. In an epistle written in 1750 he mentions again Heitman's book and puts forward an argument on the Sun's heat which he himself had argued and which he thought strengthened Heitman's ideas. Finally, as late as 1754 (i.e., 13 years after Heitman's book was published) Holberg comments again on Heitman's writing in one of his last epistles.

5.5 Ludvig Holberg Recommended Heitman's Book

A well-known learned Norwegian writer at that time, Ludvig Holberg (1684–1754) had obtained a copy of Heitman's book and studied it thoroughly. Holberg was by no means a physical scientist, but he kept up well with all that was going on at that time. He could therefore read Heitman's book with an open mind. Since Norway was a part of Denmark and Holberg had to live in Copenhagen to be in contact with the academic milieu, he in particular admired any provincial Norwegian who had courage enough to refute authorities, either on a question of politics or physics. At any rate in his epistle number 165 in 1748 he writes the following about Heitman's book:

"There are a lot of his opinions of which I cannot approve, but admit, however, that the book has generally pleased me. One finds there certain original ideas, which, if they are not properly based, give rise to a lot of reflecting and is not easy for everyone to understand. He does not completely agree with either Cartesius, Newton, Leibnitz or Wolff, but bases his opinions solely upon his own meditations and on the experiments which he himself has done".

Holberg lists four objections to Heitman's theory about sunlight, but in spite of his objections, he explains:

"However, despite this, I admit that the book had pleased me, yes even so much that I had wished for it to be translated into other languages. One sees that the author has not been a slave to the opinions of others but has sought by his own meditation and experiments to establish a system on his own".

5.6 Peter Møller's Reactions to Heitman's Book

Heitman's book stirred up strong debates. In the same year that Heitman's book was published, a navigation teacher from Trondheim named Peter Møller came out with a publication called *Impressions About the Northern Light*. It probably had been written already in 1738 but was expanded by an *Advertissement* – a short article – before it was published in 1741. In this supplement, Møller directs a sharp criticism at Heitman.

Møller had problems in getting his article printed and it could well have been a contributing factor in his lashing out at Johan Heitman. In particular he criticized Heitman Jr. who according to Møller "after his father's death, (he) in a sinister manner had let the book see the light of day".

Møller considered himself to be a somewhat better physicist than Heitman and characterized Heitman's disputes with the renowned Newton and others as a contest between a mouse and an elephant.

At the same time it is interesting to note that Møller believed the northern light which was seen in Bergen on New Year's Eve in 1702 was the very first one which was ever seen there. As Ramus had done a quarter of a century earlier, Møller also established that the northern light did obtain its light neither from the Sun nor the Moon. He contended that it was due to sulphurous vapour (or gases) which ascended from the Earth. Møller also stated that it is possible with the help of a kerchief to bring the northern light all the way down to the Earth, a well-known popular superstition in the older days (cf. Chap. 1). Variations in the occurrence frequency of northern lights puzzled him in the same way as variations in the northern light along with the weather had been a concern of Heitman. The differences in the two viewpoints, however, were not so great as to justify Møller's disparaging remarks about Heitman. Nor do they unveil such a realism that they support or substantiate Møller's concluding strong criticism about the copious summation in Heitman's' work. Møller concluded his remarks about the northern light with a warning to all sinful and irreparable people that doomsday could be near when the northern light reveals itself as if the

heavens were on fire. It can be a sign that heaven and Earth shall be burned up just as the rainbow was once made by God to be a warning to people about the flood.

5.7 Joachim Frederik Ramus – The First Norwegian Natural Scientist Who Wrote an Auroral Paper

Joachim Frederik Ramus, a nephew of Jonas Ramus, was born in Trondheim in 1685 or 1686. He was one of Norway's first professors in natural science. His father was the cartographer Melcior Ramus who died when Joachim was barely seven years old. Beginning in 1698, Joachim lived with his distinguished uncle Jonas Ramus in Norderhov (cf. Chap. 5.2) and was taught by him. After being at the University of Copenhagen for 13 years, he was appointed professor of mathematics in 1720.

Fig. 5.7. These are a few examples of curious drawings of northern lights published in one of the theses by J. F. Ramus. They illustrate the northern light seen in Danzig on the 17th March, 1716. The whole northern sky looked as if on fire. (See also Fig. 4.2)

59

Ramus received a number of positions of honor in Copenhagen and in 1742 became a member of the Royal Danish Scientific Society. In the first and third volume of the annals from this society which were printed respectively in 1745 and 1747, Ramus wrote two parts of the large publication entitled *Historical and Physical Description of the Strange Shape, Nature and Origin of the Northern Light*. The third part which he also discussed briefly and which was believed to contain his own and other people's theories about the northern light, was without doubt never finished.

In contrast to the writing of Heitman and Møller on the northern light, Ramus' work impresses one as being the composition of a learned scholar. It was written on a higher level than the controversial discussion which followed Heitman's book. These controversies are not mentioned at all by Ramus, although it is quite certain that he was aware of them.

The first part of his publication was based on an enormous amount of ancient material. Ramus tried to prove, that many of the innumerable descriptions of sundry meteors and comets, which are found in literature from the Greek philosophers and other writers up to his own time, actually dealt with the northern light. But he was very careful not to say any of the light phenomena portrayed in the Bible could have been connected to the same phenomenon. In spite of the fact that Ramus was a well-educated realist and a critical reader, he had also been reared in a very devout and religious atmosphere which would not allow him to question or doubt any part of the Holy Scripture.

Ramus had probably not read Halley's theory but extended the details of de Mairan's treatise on the northern light to a great extent. He felt that it was almost an insult that the Frenchman tried to prove that the northern light was absent in Norway between 1621–1686. This was for Ramus very close to being a shame to Norwegian national pride, and he set out in his first work to profoundly disprove de Mairan's theories. He was especially critical of de Mairan's use of one partial report from travelling in the Arctic region, which was made during the above dates. The northern light was not mentioned but de Mairan had nevertheless made use of this report to support his position. Ramus showed that the trip was made during the time of year when there was practically no darkness and the reports could therefore not contain any descriptions of the phenomenon.

Ramus attacked de Mairan's theory on a second point. He could easily prove that if de Mairan's contention that auroral heights were more than 800 km then this would imply that most northern lights seen in Scandinavia could also be seen in southern Europe. Ramus felt that this was ridiculous since experience showed that the northern light appears far more frequently in Scandinavia than any other place in Europe.

In the last part of his publication Ramus showed how the northern light earlier had been used to frighten people into obedience and to increase taxes, particularly by the reigning authorities in the Middle Ages. He mentioned as a concrete example the Roman consul Sempronius who ordered an increase in the sacrifices to the gods after a number of luminous sky signs had been seen during Hannibal's siege of Rome.

At last Ramus gave a speech defending those who were precise and unprejudiced viewers of the northern light. In general he praised all who wished to understand nature's wonders through experiments and observations and says: "how could we otherwise be saved from the many devils and ghosts that the papistic teachers by their vain philosphy have filled the air with, when we blindfolded shall believe that there are lifelike shapes and pictures, resembling human beings, animals, different sorts of armaments and other things in the air, that their tradition and writings deal with".

It is thought that the third publication was never finished because Ramus was blind for a time and lived with strongly impaired vision, until his death in 1769. It is very unfortunate for us that the third volume, which was thought to contain Ramus' own theory on the northern light, will never see the light of day. But the two volumes which he did write provided during his lifetime a great boost, which was so greatly needed to the Norwegian's national pride. They were written in a language which the native Norwegians could understand and would become solid references for Norwegian writers to use throughout the century.

5.8 Anders Celsius Set Out on a New Course in Auroral Research

Arnelius' treatise (Chap. 5.1) did not create any great enthusiasm for the work in Sweden, despite the fact that in Uppsala there were many learned scientists in

Fig. 5.8. Anders Celsius (1701–1744) professor in mathematics in Uppsala Sweden and most famous for inventing the temperature scale still carrying his name. He was awarded leave of absence in 1732 and visited several university centres in Europe. He became interested in the northern light, and after returning to Sweden he devoted part of his time to the study of this phenomenon

the beginning of the 18th century, nor was it noticed in other countries. Almost 25 years passed before another Swede bothered to discuss the phenomenon.

Anders Celsius (1701–1744), who today is best known for his temperature scale, wrote his first important treatise on the northern light in 1733. The paper created a sensation over all Europe. It was written in Latin and bore the title *CCCXVI Observationes of Lumine Boreali, ab a MDCCXVI at. a MDCCXXXII partim a de, par tim ab allis, in Svecia habitas* or in English *316 Observations of the Northern Light made in Sweden from 1716–1732.*

The writing was inspired by a similar work published by a German named J.F. Weidler, and it indicated that Celsius' stay abroad must have developed in him a more basic scientific discipline than was evident from the entry in his diary which was made when he left Sweden the year before (Chap. 1.1).

Celsius spent some time on the many fantastic and ingenious explanations of the northern light which had been proposed in his time, but noted ironically that "it is deplorable that the northern light still has not let itself be tamed by the scientific community".

In this manner he emphasized that all speculations should be distrusted, and reminded his contemporary scholars that to understand true natural phenomena they should forget about speculations, and rather endeavour to make accurate observations, and exact reports on these observations. Celsius so rightly maintained that only when the northern light is observed from different positions simultaneously can one hope to attain data which can bring knowledge about the phenomenon a step forward. His list of observations is an excellent example of one such effort.

Celsius had little confidence that his contemporaries would be able to unravel the northern light riddle, but he did not worry about it. He thought that future scientists would commend his colleagues far more for accurate recordings of observed northern lights than on wild speculations. Celsius felt that the most important function of science was to serve mankind and not to glorify scientists. Therefore, there should not be any reason to regret later the possibility to have reaped a harvest from his present work.

Celsius' treatise reminds one of a lecture on ethics to people writing a scientific paper. It is easy to understand that it received much attention in Europe where many writers had contributed one speculation after another on the northern light without, perhaps, ever having seen the phenomenon.

For Celsius the northern light remained a phenomenon which occurs in the lower atmosphere. He did not provide any independent theory on the northern light but had little faith in de Mairan's explanation.

Celsius also criticised Gasendi (Chap. 4.3) and other European physicists for inventing the Latin name *aurora borealis* on a phenomenon which for generations among the Scandinavians had been named the *northern light* (Chap. 2.2). Celsius therefore demonstrated against this European lack of tradition and used his own Latin translation of the word, namely *lumine boreali.*

61

5.9 Olof Peter Hiorter – The Northern Light and the Compass Needle

The interest in the northern light in Europe was mainly awakened with de Mairan's book. Celsius' treatise, however, gave the group in Uppsala a reputation which endured until the end of the 18th century.

Celsius had got his close colleague and brother-in-law, Olof Peter Hiorter (1696–1750), interested in the northern light and especially had opened his eyes to a possible correlation between the northern light and variations in the position of the compass needle. Hiorter, who was a thorough observer, received a magnetic needle from Celsius and was very fascinated. He placed the needle on a table in his room and watched it day and night, hour by hour. For one whole year from the 19th of January 1741 until the 19th of January 1742 he took altogether 6,638 hourly readings of the needle's position (the year has 8,760 hours). This arduous and painstaking task was only interrupted by a trip home between the 13th of August and the 15th of September and a 10-day Christmas vacation. It was an imposing and dedicated task which exemplifies a unique self-discipline and work habit at that time in history.

In his treatise Hiorter expresses himself in the following manner as he describes the discovery of the correlation between the northern light and the magnetic needle's movement: "But who would have been able to predict that the northern light had a correlation with the magnetic needle?".

Hiorter had a deep sense of gratitude to Celsius and bestowed upon him, in broad outline, great honour for that work, but there is hardly any doubt that Hiorter himself deserves the distinction.

Hiorter emphasized the necessity to be cautious and to take auroral disturbances into account when the compass needle is used for navigating, surveying, and metal seeking.

He pointed out that when prospecting for metal one can obtain incorrect compass measurements which can be of great economical importance to mining companies. For that purpose Hiorter recommended that one should have two compasses; one firmly mounted in its own casing and under regular watch, while the other should be placed near the ground where the occurrence of metal is being explored.

Hiorter, who was a stimulating lecturer, awakened an interest in Pehr Wilhelm Wargentin (1717–1783) in the field of astronomy. Wargentin was interested in the history of the northern light and believed that the first northern light over Italy was observed in 1722. The most important contribution which he left for later generations in the research of the northern light, however, was the discovery that the northern light can be visible in North Scandinavia, North Asia, and North America at the same time. This was a fact which led to the conclusion that the northern light is located in a luminous ring in the air encircling the North Pole. Here he touched on a theme which two hundred years later would be a subject for an intense analysis, namely the auroral oval.

But it was really his colleague Pehr Kalm (1716–1779), who had made this discovery possible. Kalm was appointed professor in economy and sent to North America on a research expedition. According to his instructions he should go to a country with a climate which resembled that of Sweden and gather seeds and plants which he found and bring them

Fig. 5.9. Per Wilhelm Wargentin (1717–1783) studied astronomy in Uppsala where Celsius was his teacher. He studied the eclipses of the Jovian moons and made a table of these which became widely distributed and accepted. He probably was the first to mention that the northern light forms a ring around the northern pole

back to the homeland. While he was in North America he was keenly interested in everything happening in science. He noted the frequency and the location of the northern light which he observed. These Wargentin later compared with simultaneous Swedish auroral observations which finally led him to the previously mentioned conclusion.

5.10 The First "Artificial Northern Light" was Produced by a Poet

In Kongl. Vetenskap Academiens Handlinger, (The Proceedings of the Royal Swedish Academy of Science and Letters) in Uppsala one of the first spires of auroral experiment was presented. The experiment by the poet and diplomat Samuel von Triewald (1688–1742) was very primitive. The account is given here for curiosity's sake only. In one place indeed experimental physics must start; and in the context of the northern light there was the poet Triewald or "the Swedish Boileau" by which he is also called, who ignited the first artificial northern light, so to say.

His arrangement was a glass of cognac, a prism, and a screen placed in a dark room with a little hole

Fig. 5.10. Samuel von Triewald (1688–1742) probably the first ever to attempt to create an artificial northern light

Fig. 5.11. The experimental set up of Triewald's attempt to create northern light artificially. The solar rays enter the room through a hole in the wall to the right (*S*). The rays pass through a prism (*P*) and a glass of cognac (*G*). On the screen (*T*) to the left he saw "The most beautiful possible to bring out in a dark room". (cf. the text)

in the wall. When the Sun's rays passed through the hole in the wall they entered the prism and were refracted. The prism was placed in such a way that the refracted light rays fell parallel to the surface of the cognac and such that the light precisely skimmed the surface. So he projected the light against the screen "and then" says von Triewald,

> "one was surprised to see a naturally occurring northern light on the screen that nothing could more resemble. As the cognac surface was warmed up by the coloured Sun rays it began to evaporate, and with that comes into existence a wonderful movement on the screen ..., in which man sees all the phenomena like any natural northern light produces".

So he concluded his presentation of the experiment with an outflow of words which revealed some of the poet in him:

> "Never be man tired regardless of how long he looks at this experiment, for in addition it is by far the most beautiful one can produce in a dark room".

The experiment was considered to be so important that it was given a place in the venerable Proceedings of the Royal Swedish Academy of Science and Letters for the year 1744, and published two years after the poet's death. The reason why this glorious discussion by von Triewald became famous was that the experiment in many ways supports the most agreed-upon interpretation of the production of a northern light – namely that sunlight is refracted by diffuse gases evaporated from the Earth, and when these gases are set in motion by the wind, the northern light itself is set in motion.

5.11 Lars Barhow – An Active Observer of the Northern Light

In 1751 a paper was published in German with the title *Richtig angestellte und aufrichtig mitgetheilte Observationes von Nordlicht*. The author was Lars Barhow (1707–1754) – a parish priest in the Ørland clerical district of Norway. Barhow originally desired to publish his work in Danish, but when he tried to get the manuscript printed in his native language he ran into so much trouble that he gave up the idea and turned over the manuscript to his learned brother, Hans Barhow. His brother, having a high reputation in foreign countries, had the book translated and printed in Leipzig. The book was very well received

and quite often referred to up to the beginning of this century (Fig. 5.12).

In addition to the usual northern light often seen high in the sky, Barhow said that inhabitants of Nordkapp and East Finmark have always seen a blue light or glitter on the northwestern horizon. This blue light which remains quiet was called by the inhabitants "the true northern light" whereas the light which is located higher in the sky was called "weather light".

This account reminds us of similar phrases and descriptions used by de Mairan – namely that people living in the more northerly places can see another northern light which is located lower in the sky and whose source is not the same as for the northern light occurring at lower latitudes. At this time another distinguished physicist, a German named Wolff, had also observed these lights in the north and postulated that they were the source of the more southerly occurring northern lights. Barhow did not elaborate further on the question. He stated, however, that the northern light was observed more often after 1760 in Trondheim than in earlier times, and maintained that the northern light was a relatively new phenomenon. This question appear to have been a matter of dispute several times in the 18th century. It was not only a question of physical interest but also flavoured with national pride.

Barhow argued convincingly against the great height (about 800 km) that de Mairan had assigned

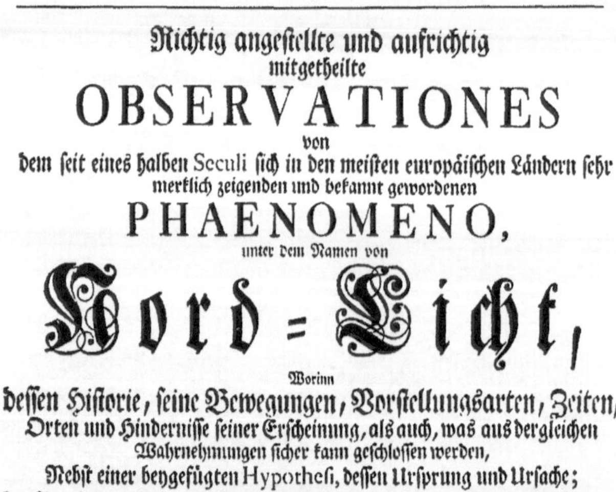

Fig. 5.12. The front page of the book on the northern light by L. Barhow which was published in 1751

Richtig angestellte und aufrichtig mitgetheilte

OBSERVATIONES

von

dem seit eines halben Seculi sich in den meisten europäischen Ländern sehr merklich zeigenden und bekannt gewordenen

PHAENOMENO,

unter dem Namen von

Nord = Licht,

Worinn

dessen Historie, seine Bewegungen, Vorstellungsarten, Zeiten, Orten und Hindernisse seiner Erscheinung, als auch, was aus dergleichen Wahrnehmungen sicher kann geschlossen werden,

Nebst einer beygefügten Hypothesi, dessen Ursprung und Ursache;

Zum Vergnügen aller Liebhaber der Naturwissenschaft, insonderheit aber zum Dienst der Physicorum, denen es an zulänglichen und in den nordlichsten Ländern genommenen Observationes bishero gemangelt hat, um den wahren Grund desselben ausfündig zu machen.

Dargelegt und herausgegeben

von

L. BARHOW,

Pastor auf Oerelland, ohnweit Drontheim in Norwegen.

Frankfurt und Leipzig, verlegts Franz Christian Mumme. 1751.

Fig. 5.13. Barhow distinguished between three main groups of the northern light. The first is an arc (*A*) which can be 30°–40° in the east/west direction, situated on the northern horizon, and moving slowly towards zenith. The second group is very faint and almost colourless marked *B* in the figure. The third group has such a manifold of colours, forms and movements that Barhow was unable to describe them in details (marked *C* and *D*)

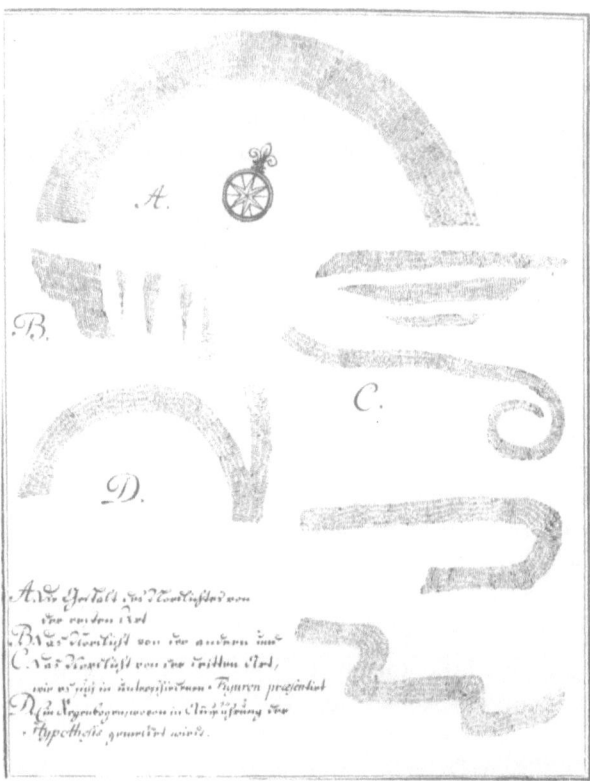

to the northern light. He pointed out that if the northern light occurs at such an altitude high above the Earth's atmosphere, then it should move in the same direction as the stars (that is from East to West) but since this is not always true, then the altitude must be less than that which de Mairan maintained.

Barhow believed that one could not determine the height of the northern light by triangulation from two places at great distances apart, as de Mairan had done, because of the small probability that one could see the same northern light from the two stations simultaneously. On the other hand if, as some authors claimed, the northern light does actually occur at such a low altitude in the atmosphere that it blows around people's faces, shakes mountain tops, or is lower than the clouds, then its motion should be in the same direction as the clouds. This did not agree with Barhow's observations. He had seen the northern light move towards the clouds but had never seen it as low as the highest mountain and neither had he observed the clouds through the northern light. Therefore, Barhow came to the conclusion that the northern light must occur above the clouds, in the uppermost layers of the atmosphere.

Barhow's other ideas concerning the northern light may be summarized as follows:

1. Matter from which the northern light originates is not self-luminous.
2. It does not consist of phosphoric or electrical matter.
3. There is a steamy vapour in the atmosphere which is illuminated from an outside source.
4. This illumination does not come from the Sun, Moon, or stars but from ice which is found around the north pole.

It is evident that Barhow correctly analyzed what determines the height of the northern light, but regarding the timely question of where the lights originate, he was in close agreement with the third explanation in *The King's Mirror* (Chap. 2.2).

Although his book was widely read, it really did not shed any new light on this matter. His auroral theory was the one most widely accepted at that time

and has been strong enough to persist, in certain circles, up to the present time. It was a simple and easily understood theory which Barhow advocated, and it was attractive to people who did not have the time or the ability to delve deeper into the problem but who only asked for a theory that gave an easily intelligible explanation of the phenomenon. What was more natural than that snow and ice, which also belong to the polar region where the northern light is primarily seen, play an essential role in the production of the phenomenon?

Barhow's greatest contribution was his attempt to classify the northern light into shapes, movements, and colour. This was an investigation started by Barhow which has kept auroral scientists busy up to the present time. He pointed out that dissimilar types of northern lights could all have the same source since the shape could change itself into the second and third principal types.

He distinguished two types of auroral motion – an interior and an exterior one with the former being more active. Regarding the arc movement from north to south, according to Barhow there is an outward motion which is always directed straight against the wind. Movement along the arc is classified as interior motion – not of mass per se, but rather a wave motion.

It was well known that Barhow was blessed with unusually good vision and hearing and the fact that he had never heard any sound from the northern light was used as an argument against the existence of auroral sound (Chap. 8.6) for several years after his death.

It is certain that Barhow must have been a very careful and accurate observer – in a class with the best astronomers. How difficult it must have been to bring order into the various and sundry auroral observations, or to record a phenomenon which seemed to defy all logic in contrast with movements of the stars, planets, and comets in their unvarying geometric paths. At any rate, the Ørland priest pointed out, for the first time, characteristics of the form and motion of the northern light that, thanks to careful techniques, now in modern times are close to a solution.

5.12 Erich Pontoppidan – The Northern Light is "an Electrical Phenomenon"

There were many who accepted Barhow's hypotheses but there were also those who perceived weaknesses in them. In this latter group the most fa-mous was Erich Pontoppidan (1698–1764) who was doctor of theology and the Bishop of Bergen. His large work *The First Essay on Norway's Natural History* (1751–1753) was a book in which he covered practically everything from oil in the North Sea to Norway's snakes and insects. The chapter devoted to the northern light begins with a quotation from Johan Heitman (Chap. 5.4) which talks about "a well-experienced and well-read Philosphers Practicus". In particular, Pontoppidan thought that Heitman's conjectures about the cause of the northern light must be capable of withstanding attacks from many other speculations. In this he had precisely the same idea as Holberg (Chap. 5.5). He agreed with Heitman's speculations which tended to substantiate the universal interpretation of the northern light as some kind of lightning phenomenon – similar to that which had been advocated by Anaxagoras.

Fig. 5.14. a Erich Pontoppidan d.y. (1698–1764) is best known for his explanation of the *Lutheran catechismus* and his book of psalms. From 1738 he was an extraordinary professor in theology. In 1746 he became bishop of Bergen. *b* The front page of Pontoppidan's book (cf. Fig. 5.15)

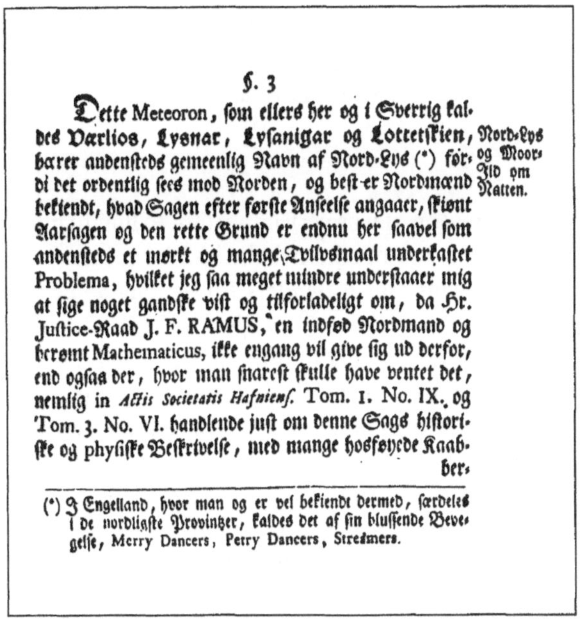

§. 3

Dette Meteoron, som ellers her og i Sverrig kaldes Værlios, Lysnar, Lysanigar og Lottetskien, Nord-Lys bærer andensteds gemeenlig Navn af Nord-Lys (*) fordi det ordentlig sees mod Norden, og best er Nordmænd bekiendt, hvad Sagen efter første Anseelse angaaer, skiønt Aarsagen og den rette Grund er endnu her saavel som andensteds et mørkt og mange Tvilvsmaal underkastet Problema, hvilket jeg saa meget mindre understaaer mig at sige noget gandske vist og tilforladeligt om, da Hr. Justice-Raad J. F. RAMUS, en indfød Nordmand og berømt Mathematicus, ikke engang vil give sig ud derfor, end ogsaa der, hvor man snarest skulle have ventet det, nemlig in *Actis Societatis Hafnienf.* Tom. I. No. IX. og Tom. 3. No. VI. handlende just om denne Sags historiske og physiske Beskrivelse, med mange hosføyede Raaber.

(*) I Engelland, hvor man og er vel bekiendt dermed, særdeles i de nordligste Provintzer, kaldes det af sin blussende Bevegelse, Merry Dancers, Petry Dancers, Stredmers.

Fig. 5.15. Erich Pontoppidan's contribution to the Norwegian history on natural science is well appreciated. Here is a page of his book *The First Essay on Norway's Natural History* (1751–1753) where he devotes a whole section to the northern light, and relates it to electricity

ence of these watery vapours on the air is the same as recently has been described concerning the influence on the glass ball by the penetrating air. Namely, this air forces the lighter *aether* air to move upwards, and is seen there for a short while like the purple fluid outside the glass ball, until it is mixed and dissipates. Just as this electrical phenomenon depends on the rotation speed of the glass ball, the northern light will also occur at the poles, because these are closest to the rotation axes. Here they have higher velocities as the air at the poles is heavier and the pressure stronger since it is forced back by less power than on the middle of the globe where the centrifugal force is directed radially outwards".

We notice that Pontoppidan mentions both poles and did not make any difference with respect to the polar light at the two hemispheres. Furthermore he had also a relatively good understanding of the variation of the centrifugal force with latitude.

Thus, Pontoppidan, the theological doctor and Bishop of Bergen, had more capabilities than merely reciting the Lord's Prayer, and his open mindedness for every happening in his time, must have been rather unique in Norway. Since men strongly trained in pious thinking dominated the current way of life in Pontoppidan's age, he must have had an inner conviction strong enough to break away, in his thinking, from the biased opinions that prevented free thinking.

Pontoppidan's auroral theory was far ahead of his time, and he is believed to have been the first who showed a relationship between the northern light and electrical phenomena. The lengthy speculation about the "aetheric air" is not valid. His idea that the earth's rotation plays a primary role in creating northern light was introduced again in the following century with a theory about unipolar induction (Chap. 6.4) – which was actually accepted among scientists up to the beginning of the current century (i.e., the 20th).

Pontoppidan was leery about the speculation that volcanic mountains were the cause of the northern light and he also disagreed with the broad interpretation that the northern light was merely the result of reflected and dispersed solar rays.

Pontoppidan believed that the northern light was an electrical phenomenon. Through a friend, he became aware of the work done by the Frenchman, Desaguliers, who had used an electrical voltage to keep heavy raindrops airborne. Pontoppidan had a chance to read this book and it inspired him to search for a theory on the northern light.

The following is an example of how Pontoppidan's reasoning was used on the subject:

"One can imagine the Earth's globe with its air-circle to be like a glass ball on an electric machine. When the air is pumped out and the ball is rapidly rotated, a purple coloured flame is produced in the air, the same colour which the northern light also posesses. This flame must be *Aether igneus*. If now the thick air is let in again, then the illuminated fluid or aetheric air will be forced out of the glass bulb. The fire will collect on the surface of the ball for a while until it is dissipated and damped by the mixture with the air. Maybe this could lead us to think the following: this northern light which is seen toward the pole, or the Earth's axis, is not only caused by aetheric air but is aetheric air, is forced by the penetrating watery vapours to fly upwards and remain almost swimming on the top of the clouds. The motion of the clouds, however, also makes the northern light unsteady. As long as the air remains dry, either in the winter when it is very cold or in the summer when it is very warm, the northern light is never seen. When the weather is changing, however, either such that mild weather follows a period of strong cold or very hot weather is followed by rain, such that some of the watery vapours are becoming present in advance, then the northern light is seen as a certain omen of variable weather. In such a situation the influ-

5.13 Gerhard Schøning Compiled Literature of the Northern Light

It was not an easy task for anyone to follow in the footsteps of such a genius as Pontoppidan. In his 1760 publication *The Age of the Northern Light is*

Fig. 5.16. Gerhard Schøning (1722–1780) was a well-known Norwegian historian, and rector of the Trondheim cathedral school from 1757. He was one of the founders of the Royal Norwegian Company of Science in Trondheim

For this reason, it was very important to him to make it clear that the northern light had been known to Scandinavians and Icelanders for as long as people had lived in these areas. He wished to refute the claim by Barhow and others that the northern light was a new phenomenon which had only become well known in the beginning of the 19th century. Barhow maintained, and Schøning agreed, that the northern light is due to the Sun's rays being reflected from the ice around the area of Greenland. Then assuming this to be true, if the northern light had never been observed in Norway in the past, it would follow that ice had never been found around Greenland and to Schøning this was preposterous.

Schøning's main idea can be summarized as follows: Even though a strong case had not been made for eternal existence of the northern light in Norway, the evidence did not prove the opposite – namely that the northern light had never existed. Therefore it was mandatory to gather as much data as possible to show that the northern light did occur in past times.

Schøning was exceptionally well read and meticulously searched through original published works on the northern light. This is clearly demonstrated by the innumerable quotations – for the most part in Latin – which he had placed in chronological order from a few hundred years before Christ up to his own time.

Schøning's overpowering desire to prove the eternal existence of the northern light in Norway strongly influenced his ability to be objective in his critical evaluation of previously described heavenly phenomena. For this reason, many of his listings of northern lights also included other light phenomena.

No new deeper understanding of the physical cause of the northern light is discussed, but his work represents a classical example of how a scientist in his writings attacks a hypothesis. In its outlines Schøning's treatise set a pattern for many educated Scandinavians. But perhaps the most important thing Schøning did for Norway's national pride was to humble the previous Danish influence. The northern light belonged to Norway, and no foreigner should say that the ancestors of Norwegians were not at home with it.

Proved with Ancient Writers' Testimony Gerhard Schøning (1722–1780) was not able to approach the bishop's physical insight into the problem. His treatise, which was published by the Royal Danish Scientific Society, was primarily a detailed bibliography of historical auroral data which did not contribute nearly so much original work as had J.F. Ramus in his earlier publications. Compared with Ramus (who, it should be noted, never wrote down his theory on the northern light), Schøning in the last part of his treatise dealt with speculations on the source of northern light. He began by referring to that passage in Barhow's book where the following matter was discussed:

"The writing by no means dishonours its author, but corresponds very well to its title, and is without doubt the best of its kind which so far has been seen" and further "necessarily it must be appreciated by most people, and very well completed with many other so-called *truths* which have been presented as completely proved during our time. Foreigners who have such low ideas about our countries Scholars, can learn from this work that the Norwegian Cliffs are not, as they think, so empty of learned people".

Schøning was one of the founders of the Royal Norwegian Scientific Society in Trondheim and to him the northern light was a thing of national pride.

5.14 Erich Johan Jessen-Schardebøll Describes the Geographic Location of the Northern Light

In 1763 a Dane named Erich Johan Jessen Schardebøll (1705–1783) wrote a paper entitled *The Kingdom of Norway Displayed from its Natural and Civil State*. In 1743 he had begun to compose another paper called *Description of the Royal Empire and Countries*. In this manuscript there were nearly a hundred closely written pages concerning the northern light and Jessen apologized for the lengthiness of his writing with the following:

"... but since this is about Norway, where the northern light for the most part has its home, and those who do not understand different languages wish to know a little more about it than what is written in our own, it should serve as an excuse, if this chapter is a little lengthy".

Jessen set out to explain the northern light in such a way that the everyday reader would understand it, and he therefore described in detail all the observations and theories with which he was familiar. He made an elegant argument against Halley's magnetic theory and all other speculative electrical theories. He made it distinctly clear that Barhow's theory was one of the most attractive of the lot, and he expressed a deep admiration for this provincial expert on the northern light whose education was in the clergy.

Jessen began to ponder over the "genuine northern light" in the north which de Mairan and Wolff had claimed to be the source for the other northern light. But he suggested that this bluish light, which was also well known in Greenland, received its reflected light from ice or water. He also thought that this light was that which oldtimers had concluded was an eternally burning fire around the North Pole. He ended his description of this light with the following words:

"It does not rise like a monument, but stands like a little arch or ball, that is seen in the sky even though above it may be seen the flickering and moving northern light".

Barhow began his work by classifying the northern light in regard to form, colour and motion. Jessen expanded this work by dividing the northern light into seven categories, compared to Barhows three, plus "the actual northern light in Scandinavia" as he referred to it. The northern light which Jessen placed in the fifth category is a bright red light which covers practically the whole sky. The ancient people used this as an omen for bad times to come. Jessen's remarks about this were:

"So much is to be admitted, that of the thoughts and ideas that common people have, these are quite reasonable."

Jessen was therefore not quite free to believe that the northern light possessed predictive characteristics.

As to the geographic position of the northern light, he said that it formed a ring around the pole which necessarily is quite continuous, and in all cases interrupted where the sun is bright. Therefore, one sees an arc stretching from the east horizon to the west horizon. These need not be the ends of the arc for it can be continuous below the horizon all the way around the pole. He wondered whether similar "weather lights" might not also be seen around the South Pole. At any rate, Jessen thought that there were good reasons for believing this, but he had only scattered rumours from "far travelling seaman" on which to base his opinion.

Fig. 5.17. Erich Johan Jessen Schardebøll (1705–1783) devoted himself to making the knowledge of the northern light understandable to the people. He also claimed that the northern light forms a ring around the pole

Fig. 5.18. Torben Bergman (1735–1784), Swedish mathematician and astronomer. He determined the height of altogether 11 northern lights

ral sounds to reach the ground. For the same reason he also seriously doubted that the northern light could have any connection with the weather.

Bergman is a good example of the topnotch physicist with the deepest understanding of auroral phenomenon. His arguments against Barhow's theory are so fundamental and well grounded that it is rather odd that Barhow's book was used for more than a 100 years after the publication of Bergman's thesis.

5.16 Johann Carl Wilcke – The Center of an Auroral Corona is the Axis of Its Magnetic Field Line

In 1768 a dissertation was brought forth, by Johann Carl Wilcke (1732–1796) from Sweden, by the Royal Academy of Sciences and Letters which had the title *Experiments to Produce a Magnetic Inclination Chart*. In this dissertation, a chart was included which showed how a magnetic needle will position itself in a vertical plane at different places on the Earth's surface. The position of the compass needle is given by the angle (inclination) the needle makes with a vertical line at the place of interest. This demonstrated that places with the same magnetic inclination lie almost in concentric circular lines around the Earth. The centre of these concentric lines does not fall at the geographic North Pole, but is displaced some distance toward the west, almost at Baffin Bay in Canada. At that time this chart represented a high point in the Scandinavian tradition concerning research output on the Earth's magnetic field. This began with Celsius and Hiorter and then continued with Hansteen (Chap. 6.1).

With his work on the inclination chart Wilcke created great interest in the daily variations of the Earth's magnetic field. As Hiorter had previously done, he installed a magnetic compass in his study room. However, Wilcke did not show the same dedication to his experiments as Hiorter had done, but noted precisely all the phenomena which could have a bearing on the reading of the compass needle. In 1777 he was able to present his new results to the Academy and he confirmed Hiorter's report (Chap. 5.9) that the point of the compass needle

5.15 Torben Bergman Measured the Height of the Northern Light

On 17th October 1763 a northern light occurred at Uppsala, Sweden. An astronomer, Fredric Mallet (1728–1797), concluded from observations at this time that the light did not always occur farther away than the clouds. Mallet had watched the phenomenon closely and thought he saw the clouds through the northern light on several occasions.

In the Proceedings from the Royal Swedish Academy of Science and Letters for 1764 the far more experienced mathematician and astronomer Torben Bergman (1735–1784) refuted this absurd claim. Bergman had a pyromaniac's interest in fires and perhaps considered himself more experienced with auroral flames than any other of his colleagues. He also observed the northern light on the 17th October and was quite certain that the clouds were below it.

In this paper he notes that he had made many observations to determine the height of the northern light. He used, to some extent, both parallax and angle observations on auroral arcs and found that the northern light can lie anywhere between 380 and 1,300 km above the ground. This showed that the northern light always occurs above the clouds. He claimed that since it is located at such heights, where the air is so rarefied, it would be impossible for auro-

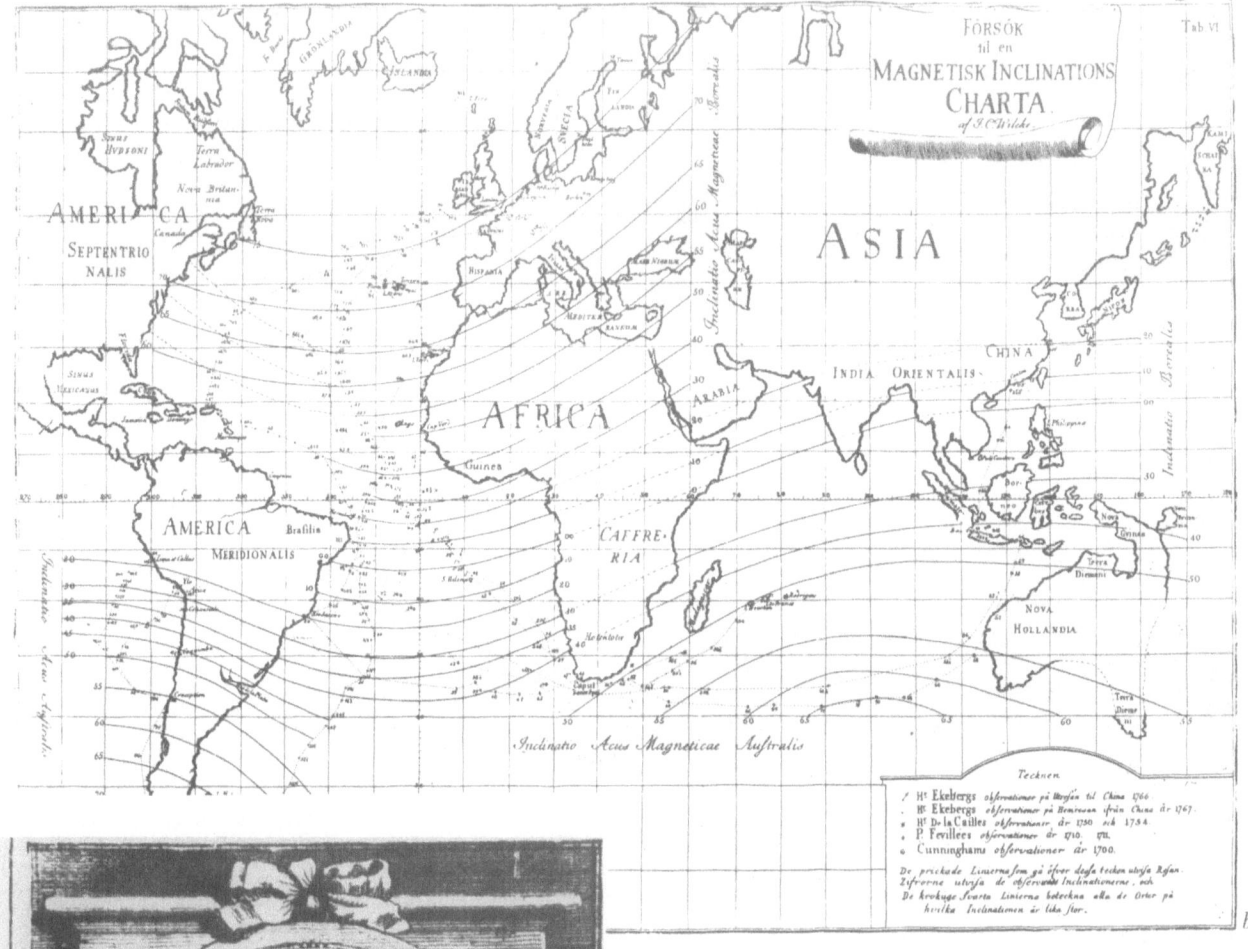

FÖRSÖK
til en
MAGNETISK INCLINATIONS
CHARTA.
af J. C. Wilcke.

Tab VI

Fig. 5.19. a Johann Carl Wilcke (1732–1796), Swedish professor in physics. He was mainly interested in electricity and thermophysics, and studied the specific heat of different materials. He also made one of the first world wide magnetic maps *b.* Wilcke's map showing the direction of the Earth's magnetic field with respect to the vertical axes at the surface (the inclination angle). Positions with equal inclination angles are situated approximately at concentric circles.

seems to follow the northern light and pulls itself towards it. For example, if the northern light was seen in the east, the needle pointed toward the east and if in the west, the needle moved in the same direction. However, Wilcke noticed that when the northern light moved upward towards the zenith, the needle did not turn upward, but instead vibrated strongly.

He especially noted that when the northern light formed a characteristic corona close to the zenith, the centre of this corona was the same place where the compass needle pointed. He was certain that the auroral corona is a perspective phenomenon in that all the rays which appear to converge toward the centre are in reality parallel to each other. Wilcke could therefore summarize as follows: "The auroral rays align themselves in the same direction as the

71

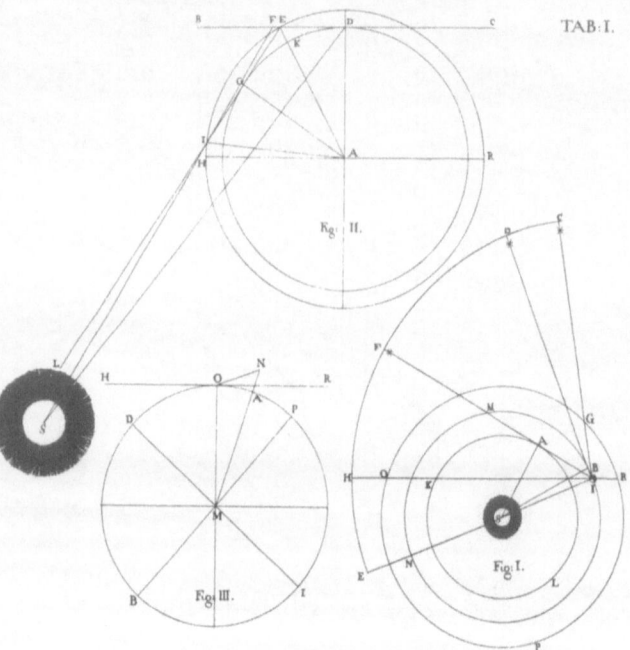

Fig. 5.20. Didrich Christian Fester (1732–1811). He was born in Denmark and self-educated in mathematics. Fester was the author of many scientific articles and among these were two concerning the northern light

Fig. 5.21. Illustration to the article by Fester from 1788 showing how to determine the height of an auroral arc N (Fig. III) and how to determine at what latitude an arc can be seen if its height is known

magnetic force which swings the inclination needle to its position".

This was powerful evidence that the northern light was not only influenced by magnetism but in fact was also oriented along the magnetic field lines. Any future auroral theory then of necessity must include processes which closely relate to this factual evidence.

5.17 Johan Ernst Gunnerus
Suggested that There Must be a Northern Light Around Venus and Mercury

It was several years before a new book on the northern light again became publicly available in Scandinavia. It happened in 1781 when the teacher in mathematics and navigation in Trondheim, Didrich Christian Fester (1732–1811) wrote the book *Mathematical and Physical Considerations about the Northern Light*. Fester, who was an autodidact, came to Trondheim from Denmark in 1769 in great measure aided by Pontoppidan's intervention with the important bishop Johan Ernst Gunnerus (1718–1773). Fester later became one of the pioneers in introducing professional education into the higher schools in Norway.

In his book on the northern light Fester opened with a survey of the theories with which he was familiar and criticized most of them. For example about Heitman's (Chap. 5.4), he said:

"By the completion of the sea-chart and other works to the benefit for navigation Mr. Heitman has real profit, but his science of nature does not agree well with his seamanship".

Regarding de Mairan and his teaching about the northern light, Fester made the following comment:

"Mr. Von Mairan's doctrine concerning the northern light will, by future scientists, be considered as wise and trustworthy as the Copernican System concerning the order in which celestial bodies move around each other".

He strongly attacked Ramus and Barhow but to an even greater extent Jessen who had dared to criticize de Mairan. On this matter Fester says that they in fact had been diligent in auroral writings but had not answered the most important question, namely what is the northern light.

Fester had an exceedingly interesting conversation with Gunnerus. During a discussion about the

Fig. 5.22. Johan Ernst Gunnerus (1718–1773). In 1758 he became bishop of Trondheim in Norway, where he also worked hard to create a scientific interest among the citizens. Together with Gerhard Schøning and others he started the Royal Norwegian Company of Science, which from 1767 became the official Royal Norwegian Academy of Science. Gunnerus was a well-known scientist, and published several articles on zoology, mineralogy, and botany

northern light Gunnerus proposed that if it is caused by particles from the solar atmosphere then it must also create northern lights around the moon, Venus and Mercury just as the light around a comet. Following this line of reasoning, the Sun must have diminished in strength since its time of creation. This thought led the author to delve deeper into mathematical considerations about the Sun's age and strength. The conclusion was that the Sun had shrunk, but it was by such a trivially small amount that there would be plenty of particles for northern lights "in an extremely high age of the world" as he expressed it.

Regarding the prediction about Venus, Mercury and other planets, it is one of the questions today which we would like to answer by sending satellites around these planets.

Fester later (1788) complemented his *Mathematical and Physical Considerations* with an article in the Proceedings from the Royal Norwegian Company of Science in Trondheim. In this paper he discussed the height of the northern light and concluded that it was close to 70 Swedish miles, that is about 700 km above the Earth. This height was a little lower than the average height derived by Bergman. It was within acceptable limits of the height derived by de Mairan, the auroral physicist he, for good reasons, admired more than anyone else.

In this last work Fester described several methods for deriving the height of the northern light (see Fig. 5.21).

Fester also tried to explain how the northern light gets its light. He, like de Mairan, believed that the cause of the northern light could be found in the contact between the Earth's upper atmosphere and the solar atmosphere. He believed that when particles in the solar atmosphere were captured by the Earth's gravitational field, they gathered in the polar regions because, as he argued: "the Earth's atmosphere is heaviest there". He also believed that these particles either were by themselves illuminating, or they could cause fire when they were mixed with particles from the Earth's atmosphere, or they could be put on fire by the influence of the air.

Finally, Fester tried to explain the observed fact that the northern lights are most frequently seen during equinoxes and more often during autumn than spring. He believed that this was due to the angle made by the Earth's rotational axes and the perpendicular to the solar equatorial plane.

5.18 Maximillian Hell's Expedition to Vardø in Norway

The Hungarian astronomer Maximillian Hell (1720–1792) was asked by the Danish-Norwegian king Christian VII to go to the fortress at Vardø on the eastern coast of Finnmark in Norway to watch the passage of Venus in 1769. In a letter to his friend Father Pilgram he said the following about the northern light:

"The northern light, this beautiful display, we observed every clear night on our travel. How many different characters! Mairan and all his supporters are enormously mistaken when they believe that it originates in the solar atmosphere. Almost any (auroral ray) except a very few are pure circular arcs, coming out of all four cardinal points of the sky, but I will leave a further description of this, the observations and my experiments to an extra section of my travelogue".

And Hell not only described the northern light but he also gave his own theory for it in *Aurorae Borealis Theoria Nova* printed in 1776 (Fig. 5.23).

While in Vardø, Hell actually tried to find a connection between the northern light and electricity. He

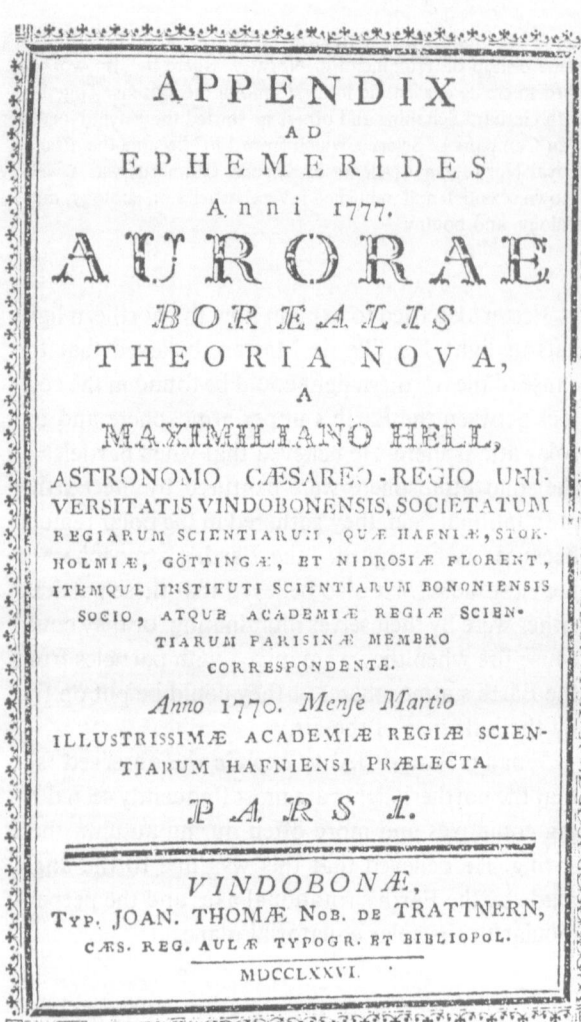

APPENDIX
AD
EPHEMERIDES
Anni 1777.
AURORAE
BOREALIS
THEORIA NOVA,
A
MAXIMILIANO HELL,
ASTRONOMO CÆSAREO-REGIO UNI-
VERSITATIS VINDOBONENSIS, SOCIETATUM
REGIARUM SCIENTIARUM, QUÆ HAFNIÆ, STOK-
HOLMIÆ, GÖTTINGÆ, ET NIDROSIÆ FLORENT,
ITEMQUE INSTITUTI SCIENTIARUM BONONIENSIS
SOCIO, ATQUE ACADEMIÆ REGIÆ SCIEN-
TIARUM PARISINÆ MEMBRO
CORRESPONDENTE.

Anno 1770. Menſe Martio
ILLUSTRISSIMÆ ACADEMIÆ REGIÆ SCIEN-
TIARUM HAFNIENSI PRÆLECTA

PARS I.

VINDOBONAE,
TYP. JOAN. THOMÆ NOB. DE TRATTNERN,
CÆS. REG. AULÆ TYPOGR. ET BIBLIOPOL.
MDCCLXXVI.

Fig. 5.23. The front page of the book by Maximilian Hell: *Aurorae Borealis Theoria Nova* which is based on his observations from Vardø in Norway

3. The northern light is an optical phenomenon caused by solar and lunar rays, usually by reflections but sometimes also by reflections and refractions together.

Hell's theory was well received in Scandinavia, probably because it was made public in Copenhagen on his way back from Vardø and also because it was based on experiments far above the polar circle in the region which was thought to be the homeland of the northern light.

Hell was probably one of the first to suggest that the occurrence of the northern light was related to the activity on the Sun. He also argued for a close relationship between the occurrence of the northern light and the occurrence of cold and windy weather. His idea was that about two months after the occurrence of a strong northern light, severe cold and windy weather would take place in the region where the northern light was seen.

In 1786 the Danish professor Christian Ulrich Detlev Eggers in his book entitled *Beschreibung von Island* (Description of Iceland) carries Hell's theory a step further and postulates that:

"This, or the substance in the air which reflects the northern light, is in reality the initial snow substance, or the finest dust of ice. It is lighter than water, and is present at higher altitudes in the atmosphere than the ascending water vapour".

In his book, however, Eggers devoted a long chapter to the northern light which represented a review of the knowledge of this phenomenon at the turn of the 18 th century. Although he was mostly in favour of Hell's theory, he realized that it could not explain all the different aspects of this mysterious phenomenon. Being a professor in political science or political economy, he uncovered insight into a very different field of science when he ended his chapter by claiming that the relation between the northern light and the weather probably will be most easily explained by the theory of electricity.

failed to show such a correlation and even could not find any influence on the magnetic needle due to northern light. He therefore felt that most of the contemporary theories were wrong, and decided to develop one of his own.

In his book he summarizes his theory into three propositions with subsequent demonstrations and colloraries as expected of a scholar of his class at that time. Hell's major statements are as follows:
1. The northern light substance is, in particular, frozen vapour of various figures and forms.
2. The radiance which can be seen in the northern light phenomenon is composed of solar and lunar rays which often have different directions.

6 Scientific Auroral Experiments Beginning in the Nineteenth Century

6.1 Christopher Hansteen and Hans Christian Ørsted Inspired Auroral Research

As the 18th century came to a close, knowledge about the northern light faced several limiting factors. First of all, the height of the northern light was in great doubt with values all the way from the treetops up to 1,300 km (above the ground). The most convincing or reliable measurements which were undertaken in Sweden, placed the northern light at a height between 350 and 1,300 km with a mean height of around 760 km (Chap. 5.15).

Many different forms of the northern light were described by Barhow (Chap. 5.11) and in particular he had observed two characteristic movements in the northern light but other than this idea, at best, the movement of the northern light could only be described as chaotic. It should be noted that Wargentin and Jessen had indicated that the northern light was located in a ring around the pole (Chaps. 5.9 and 5.14).

In 1716, the English astronomer Halley advocated a theory which related the northern light to the

Earth's magnetic field, and researchers such as Celsius and Hiorter had demonstrated that variations in the Earth's magnetic field took place when northern lights were present (Chaps. 5.8 and 5.9).

The Frenchman de Mairan had put forth a theory that the northern light was in some manner related to the Sun's atmosphere. This was a hypothesis which, if correct, implies that the northern light occurs at very high altitudes. Spidberg (Chap. 5.3) was skeptical about an altitude for the northern light as great as 800 km, because the same auroral forms were not seen over sufficiently large geographical areas. Pontoppidan and Eggers had mentioned that it could be an electrical phenomenon (Chaps. 5.12 and 5.18).

These were some of the confusing facts facing the researcher in the beginning of the 19th century. Most of the ideas were hypothetical and based on sparse observational material.

The pioneer in Norwegian natural sciences in the 19th century was Christopher Hansteen (1784–1873). He first studied law but later broke away from this profession to study mathematics and astronomy. In his student days he was warmly welcomed into the so-called Oehlenschläger Ørsted's circle where the famous Danish physicist Hans Christian Ørsted (1777–1851) awakened his interest in the Earth's magnetism.

In 1812 he entered a prize contest which asked the following question, "Can one explain all the earth's magnetic peculiarities with a single magnetic axis or does one need to assume several?" In his thesis he concluded that the Earth's magnetic field has four poles, two as we know them today. In addition he

Fig. 6.1. Christopher Hansteen (1784–1873) a Norwegian astronomer and geophysicist professor at the University of Oslo, the founder of the astronomical observatory at the university and its first director. He devoted himself to a study of geomagnetism and travelled for this reason (1828–1830) through Siberia all the way to China. The geomagnetic data collection obtained by Hansteen was an important element in Gauss's epoch-making theory of geomagnetism

the basis of this answer, Hansteen was later named professor of mathematics and astronomy at the new university in Christiania (now Oslo).

In 1829–1830, he went to Siberia to study what he believed was a pole in the Earth's magnetic field. He established several observing stations, and had arranged with sea captains of Norwegian lines to observe and record the Earth's magnetic field all over the world. On the basis of an enormous amount of data thus collected, he drew his first magnetic chart of the Earth.

Hansteen also studied the relationship between variations in the Earth's magnetic field and the northern light. He found that the horizontal component of the magnetic field increased before the northern light was seen and then decreased as soon as the light came into view.

However, his most important contribution in this connection was in proving that the highest point in an auroral arc ordinarily lies near the magnetic meridian at the location of the observation station. Hansteen explained this curving of the northern light in the form of an arc as being due to the light forming a ring around the magnetic pole at a considerable altitude above the Earth. Since he believed there were four such poles he also assumed there were four auroral rings although he was in doubt whether the ring around the Siberian pole was complete.

Hansteen believed that because the northern light formed these rings around the poles, it must originate within the poles and he, therefore, stated:

"It appears, therefore, that the polar lights spring from four points on the surface of the earth, which, so far as we have hitherto been able to determine, coincide with the magnetic poles of the Earth".

Fig.6.2. Hans Christian Ørsted (1777–1851) Danish physicist and professor at the University of Copenhagen. In 1820 he discovered a relationship between an electric current and magnetism. He also suggested that magnetic variations observed underneath an auroral arc could be explained by a discharge current along the arc

apparent center of an auroral corona, and his statement is as follows:

"and it is a very remarkable circumstance, that the distance of this corona from the southern horizon, is exactly equal to the inclination of the needle at the place; so that the south pole of the needle points directly to the centre of the corona".

He was also the first person to use the words *magnetic zenith* in this regard, which he did in the following manner:

"If we turn the eye towards the magnetic zenith (if I may be allowed to give this name to that point in the heavens to which the higher, or, with us, the Southern Pole of the needle points), we here see the luminous columns from the end".

Hansteen believed that the northern light was caused by an elastic fluid streaming out of the ground. The Earth's magnetism operated on this fluid, or portions of it, in accordance with the known laws of repulsion and attraction regardless of whether this substance was "electricity in a neutral state" or "an elementary substance, or some other state as yet unknown to us, and on which the magnetic power could act".

Hansteen thought that the auroral substance emerged from the ground and penetrated the lower atmosphere. He believed that the occurrence of the northern light had a cooling effect on the lower atmosphere which resulted in fog and cold weather. According to him:

"While the polar lights in penetrating the watery, transparent, vapours existing in the air, must have the effect, of taking from them their heat, and thereby rendering the air opaque".

And further:

"Now, as the polar light is an exhaustible substance, which, in regions surrounding the magnetic poles, continually issues from the surface of the Earth, it may be conceived, that this stream, in passing through the atmosphere, continously lowers its temperature, and thus, decomposes the watery vapours, producing fog".

Finally, he claims:

"It is a matter of experience perfectly well known here in the north, which I have found confirmed by the observations of a good many years, that the aurora borealis is generally accompanied by a strong biting cold".

This latest statement must have been in good accord with the common thought, since people in Scandinavia often believed that the northern light only occured when it was cold and clear.

In 1826, H.C. Ørsted published a *Survey on the Royal Danish Scientific Society Negotiations and its Members' Tasks*. Here he pointed out that magnetic variations in the neighbourhood of auroral arcs should be related to electric discharges along the arcs. With these few words Ørsted described the electrical current which flows in the upper atmosphere in connection with the norther light. In order to understand the ingenuity of this idea one must remember that it lay dormant for 100 years before it was fully understood within the scientific community that the polar upper atmosphere is indeed an effective current conductor.

In the beginning of the previous century interest in the northern light was practically non-existent, and there is hardly any doubt that Ørsted and Hansteen's insights in many ways rejuvenated what had become a fading interest in the northern light. That Hansteen had set himself considerable goals in auroral research becomes clear from a letter to Ørsted in 1815 in which he maintains: "undoubtedly, also a magnetic and an auroral theory should come from us".

6.2. Anders Jonas Ångstrøm, the Northern Lights Come from Luminous Gas

In order to determine whether or not the northern light is simply reflected sunlight, numerous studies were done at the beginning of the 19th century on the polarization of auroral light. Polarization is a particular property acquired by light rays after they have either been reflected or scattered. At this time there existed a so-called polarimeter which was used to study polarization properties in light rays of different kinds. In 1817 the French physicist Jean Baptist Biot (1774–1862) took a trip to Shetland to study the northern light with such a polarimeter, but his results showed no trace of polarization.

This proved that the northern light could not be reflected or scattered light rays from the sun or the moon. (This is in contrast to the situation for haloes and rainbows where these two types of scattered light rays are polarized.) The northern light must therefore be a self-luminous phenomenon.

Around the middle of the century there were several confirmations of Biot's observations, but there were also some who found weak traces of polarization in the northern light. The experimental tech-

nique was relatively primitive and therefore there were a limited number of physicists who could "ingeniously" operate an apparatus such as a polarimeter. In Scandinavia at this time the different natural sciences were at best only loosely organized and financial support, for experimental physics such as polarimetric studies was extremely limited.

At this time it was generally known in the scientific field that glowing metals or liquids emit a continuous light spectrum. A luminous gas, on the other hand, has a spectrum consisting of sharp lines and bands with dark spaces in between. The number, position and line strength are dependent upon the chemical composition of the gas.

With this background knowledge of spectroscopy at hand, a Scandinavian physicist named Anders Jonas Ångstrøm (1814–1874) proceeded to bring experimental physics up to the best international level. He began his studies of the northern light by observing the line spectra therein and he could therefore say that northern light was emitted by a luminous gas – not from a liquid, or a solid or reflected sunlight.

Fig. 6.3. Anders Jonas Ångström (1814–1874) a Swedish physicist and professor at the University of Uppsala. In 1852 he presented his absorption law, and was the first person to measure the wavelength of one line in the auroral spectrum

The line which Ångstrøm observed in his spectroscope was yellow-green in colour, and in 1868 he established its wavelength at 5,560 scale-lengths. (This unit of length later became known as the Ångstrøm (Å) 1 Å = 1/10,000,000,000 m). Ångstrøm could not identify the gas from which this line originated since it did not coincide with any line in the known spectra of simple gases or gas compounds.

6.3 Karl Selim Lemstrøm, the Northern Light was an Electric Discharge Between Earth and the Sky

Ångstrøm's work created interest among physicists in Europe and several spectroscopes were directed towards the northern light in the years which followed.

In Finland, Professor Karl Selim Lemstrøm (1838–1904) was the leader in auroral research and undertook a trip to Lappland where he industriously used his spectroscope. He thought that the auroral spectrum consisted of 12 lines which probably fell very close to well-known lines in atmospheric gases. He measured the green line's wavelength at 5,569 Ångstrøm units. The wavelength was very near

Fig.6.4. Karl Selim Lemström (1838–1904) a Finnish physicist and professor at the University of Helsinki. In 1873 he presented a thesis called: "About the electric discharge in the polar light and the polar light-spectrum". He went on several expeditions to Lappland to study the northern light in its natural environment, and he even tried to create northern light artificially in the laboratory

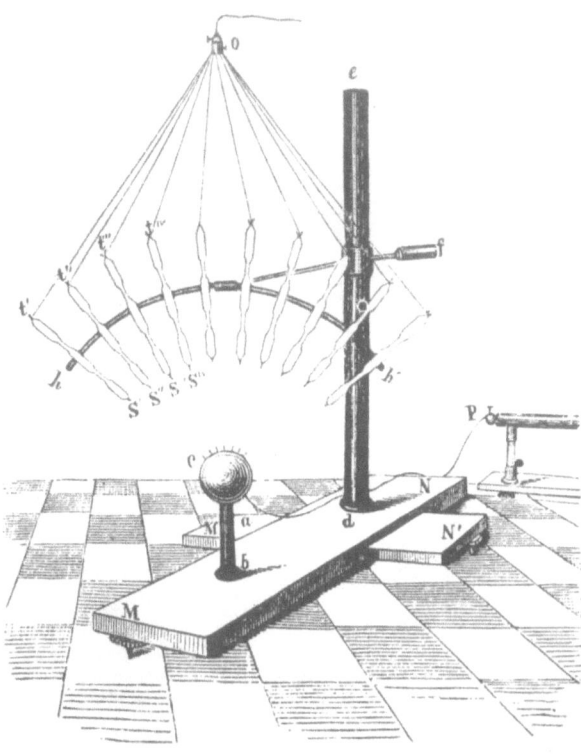

Fig.6.6. The experiment with the so-called Geisler's tubes is illustrated in this figure from Lemstrøm's book. S't', S''t'', etc. are the Geisler's tubes with reduced air pressure. C is an electrically conducting sphere furnished with pointed conductors. The sphere is connected to the negative pole (P) of the electrifying machine. All tubes point radially towards the center of the sphere and are connected by electrodes at both ends. The upper electrodes are connected to a grounded point (O). The tubes can be mounted at different distances from the sphere.

By charging the sphere negatively with an electrifying machine the tubes can light up to a distance of 2 m from the sphere if the air pressure in the tubes is low enough. When the pressure is lower than 50 mm Hg the spectrum of the light emission will be very similar to the spectrum from the northern light. Lemstrøm believed that there was not only a similarity in the two spectra but that the whole principle for formation of the northern light could be explained by this very simple physical experiment. The northern light was a discharge phenomenon between the negative Earth and the positively charged rarefied air in the upper atmosphere

a corresponding one which Lemstrøm observed when he sent an electric current through a diluted gas. He also thought that the auroral spectrum consisted of three different types, as each was dependent on which characteristic the electric discharge must have in the atmosphere.

Lemstrøm studied for a while with Professor Edlund in Sweden (Chap. 6.4) and believed along the same lines as Edlund that the northern light was an electric discharge phenomenon similar to lightning. He thought that electric discharges in the atmosphere were of two types, one fast and one slow. With the fast one he associated lightning and heat lightning, but he classified the northern light with the slow type. For Lemstrøm the northern light was an atmospheric electric phenomenon but different from lightning in that the northern light could also occur without clouds.

Fig.6.5. In his book *Om Polarljuset eller Norrskenet* (About the Polar Light or the Northern Light) in 1886, Lemstrøm presents an illustration of his discharge apparatus. His observation hut can be seen in the front of the mountain Pietavintunturi in the neighbourhood of Kultala in Finnish Lappland. On the mountain crest a few poles are mounted which are connected by an electrical conductor to a grounded plate at the foot of the hill. To the left there is an auroral arc partly hidden behind the mountain and from the hill crest an auroral ray shoots up from the discharge apparatus. All across the middle of the picture is seen another auroral-like light

For this reason the northern light could well occur over the whole earth. It was of significant importance for Lemstrøm to point out that similar phenomena occur at latitudes other than the polar region. For example he referred to light observations made along the mountain chain in Peru, Bolivia and Chile which he thought were related to the northern light.

Lemstrøm participated in the Swedish expedition to Svalbard in 1868 under the leadership of A.E. Nordenskiøld (1832–1901). When the expedition stopped in Tromsø, Norway, on its way to the Arctic, Lemstrøm went ashore with some of his instruments and made observations of the northern light. These were probably the first scientific observations of the northern light made in Tromsø, close to the site where the Auroral Observatory (Chap. 11.5) was built almost 60 years later. From Svalbard, Lemstrøm reported some special light phenomena which arose from the mountain tops and which had peculiar auroral spectroscopic properties. From his observations in Lappland he had found that the Earth's magnetic variations, which appear in connection with the northern light, could not be caused by an electric current in the atmosphere. On the contrary, he thought that it's alternation was caused by a magnetic field under the Earth's surface. For Lemstrøm this was decisive proof that his theory was correct. The northern light was an electric discharge between the earth and the atmosphere and not along the auroral arc such as Ørsted had indicated. Lemstrøm, like Hansteen, believed that the northern light emerged from the ground and penetrated the atmosphere from below. In order to prove this, in 1871 he mounted a so-called

Fig. 6.7. Erik Edlund (1819–1888) a Swedish physicist and professor. Edlund organized the first network of meteorological stations in Sweden in 1856 where a large amount of visual observations of the northern light was made

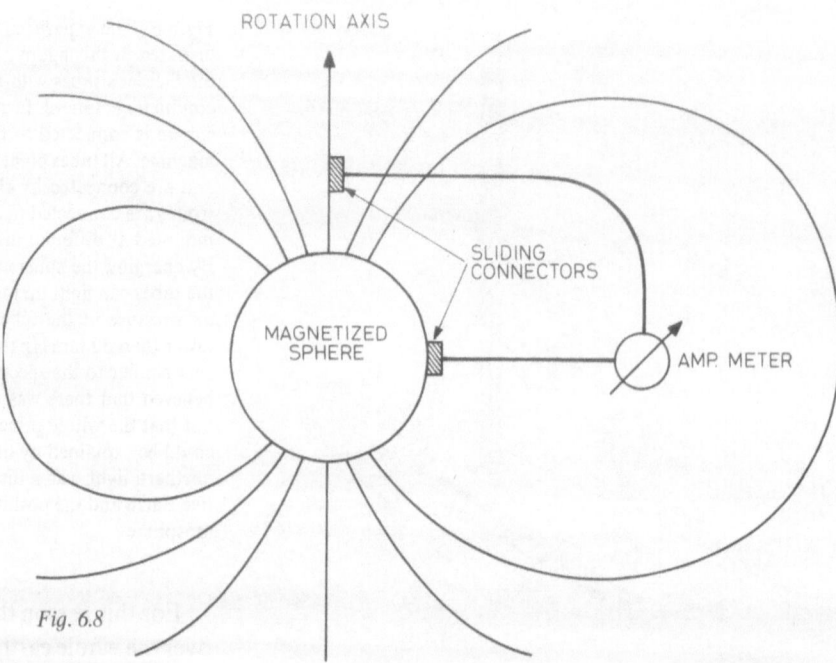

Fig. 6.8

Fig. 6.8. Edlund's auroral theory was based upon the principle of the unipolar inductor. If a conducting wire is suspended between a gliding contact on the rotation axes and at the equator of a rotating magnetized sphere, a current flows in the wire. In the same way Edlund stated that a current would flow from the equator to the poles along the magnetic field lines of the Earth

"outstreaming apparatus" on a mountain top in north Finland.

The apparatus (cf. Fig. 6.5) consisted of a ring of copper (tips or points) mounted on stakes which via isolated conductors was connected to a grounded metal plate at the foot of the mountain. The conductor contained a galvanometer in the circuit to measure the current. For several reasons he was convinced that lightning similar to northern light appeared above the discharge apparatus. The results from this experiment indicated to Lemstrøm that the northern light could be produced by the discharge apparatus and it reinforced his idea that the phenomenon had its root cause in electrical discharges from the earth.

Lemstrøm's greatest contribution to auroral research was in spectroscopic analyses. From his many measurements, taken in the field and in the laboratory, it became more convincing to researchers that the observed auroral spectral lines came from rarefied atmospheric gases at great heights. It therefore opened the possibility that with auroral research one could obtain a better understanding of the composition of the earth's upper atmosphere.

For the rest of Lemstrøm's auroral theory, one must say that he literally turned the problem upside down.

6.4 Erik Edlund's Theory of the Northern Light

The Swiss de la Rive was probably one of the key persons to advocate the theory that lightning and northern light are both caused by electrical discharges into the atmosphere. Since thunderstorms occur frequently at the lower latitudes, he called lightning in these regions "equatorial lights".

He thought electrical discharges were stronger in the polar atmosphere than elsewhere and this correlated with his idea that air moisture was at its greatest height in the polar region. This theory was well received by many people because of the simple manner in which it connected the mystical northern light with the more familiarly known lightning.

The theory, though, was objectionable because it did not give a satisfactory explanation of electrical voltages in the atmosphere and the difference between "polar light" and "equatorial light".

In 1878 a Swedish physicist Erik Edlund (1819–1888) came out with an auroral theory based on experiments. His experimental set-up is shown schematically in Fig. 6.8. In the center is a magnetized sphere to which slipping contacts are made at the pole and the equator. An ammeter is then connected in series – completing the curcuit. When the

Fig.6.9. Sophus Tromholt was born in 1851 in Husum, North Germany (then Danish), and educated to be a school teacher. Tromholt was immensely interested in science, particularly in astronomy and meteorology and with an intellectual bent which he inherited from his father. In 1875, he came as a teacher to Bergen, Norway, a position he gave up in 1882 to devote himself completely to auroral research. In order to complement his studies with personal observations of the northern light, he spent the winter 1882–1883 in Kantokeino, Finnmark and 1883–1884 in Reykjavik, Iceland. He died in 1896 in Blankenheim, Thüringen

sphere is rotated, a current is observed in the ammeter – the direction of the current and its magnitude are found to be dependent upon the sphere's direction of rotation and its rotation speed, respectively. Applying these findings to the Earth, it being a rotating magnet and the upper strata of the Earth's air being a good electrical conductor, then a current discharge path between the polar and equatorial regions should be established.

Edlund thought that this current in the Earth's atmosphere should be stronger closer to the pole. Also, at the equator where the current flowing away from it is weaker, there should exist such a strong electric field that electric discharges would occur and result in thunderstorms. Because the current would tend to follow the magnetic field lines it would also create discharges in the auroral zone. Based on this theory Edlund believed that this explained the relationship between auroral displays and disturbances in the Earth's magnetic field.

Edlund's theory was one of the most advanced at the end of the 19th century and was received very favourably by fellow workers in the field. Its soundness was based on his experimental results with the magnetized sphere, and the theory gained respect for research investigations of a type which was based on experiment rather than speculation.

6.5 Sophus Tromholt Investigated
the Periodic Occurrence of the Northern Light

Sophus Tromholt (1851–1896) from Denmark worked full time on the auroral problem and organized a net of stations ideally situated to make visual auroral observations. During his 15 years stay in Norway he made regular observations himself and noted auroral descriptions from more than 2,000 people in Scandinavia.

Beginning in 1882, Tromholt was awarded stipends from both Denmark and Norway so that he could devote full time to his great interest – auroral research. In the First International Polar Year (1882–1883) he installed a research station in Kautokeino, Finnmark, Norway. His main idea for doing so was to triangulate the height of the northern light in coordination with the official Norwegian station in Alta and the Finish station in Sodankylä. One of the reasons for establishing the First International Polar Year by people all over the world was to carry out auroral and earthly magnetic observations along the northern hemispheric auroral zone. Tromholt also organized auroral expeditions both to Iceland and North Scandinavia.

Around 1850 it was known that auroral occurrences varied in time with variations in sunspot activity. Tromholt was particularly interested in how well auroral occurrences might correlate with the 11-year

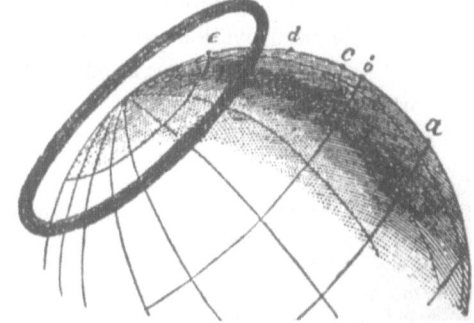

Fig.6.10. Tromholt's illustration from 1885 of the Auroral Oval

81

sunspot cycle. For the first time enough reliable data had been accumulated to make valid statistical conclusions about the correlation, if it existed. Tromholt's results are shown for the time period 1780–1877 in Fig. 2.8. From the figure it is evident that the northern light occurs more frequently in Norway during high than during low sunspot activity. Auroral occurrences in connection with variations in the Earth's magnetic field were of great interest to him and he systematically investigated why the northern lights brought interruptions to telegraph lines and particularly in the auroral zone.

Tromholt also pointed out that the northern light forms a luminous ring arround the North Pole, and he made a drawing in his book to illustrate this point (Fig. 6.10). This is one of the first illustrations of what later has become known as the *auroral oval*. Tromholt also explained the movements of the auroral forms by contraction and expansion of this luminous ring.

Today it is probably accurate to say that Tromholt is best remembered for his written auroral descriptions. He wrote easily and vividly and was a very active and popular author. A list of Tromholt's popular newspaper and timely written articles fill more than seven pages in Halvorsen's *Author-Lexicon*. His main work *Under the Rays of Aurora Borealis* was both sober and entertaining. It was first published in English in 1883 and then translated into several other languages. Tromholt was certainly a light-hearted person. For example, he wrote the following:

> "life everywhere in the world has more light than dark sides and I would not mind it a bit if my years of living could be anywhere close to Methuselah's".

Sadly though, Tromholt was only 45 years of age when he died.

During the last 10 years of his life Tromholt went on several extensive lecture tours, both in Europe and America. There is hardly any other person who has given so many popular lectures on the northern light as Tromholt. His imposing contributions have unfortunately been overlooked in modern Norwegian and international articles about the history of the northern light.

Fig.6.11. The American professor Elias Loomis (1811–1889) showed in 1860 that there is a zone around the northern pole in which the occurrence of northern lights is most frequent. This zone is now called the Auroral Zone and passes through the northern part of Norway, Iceland, and the Southern tip of Greenland. One also notices that the auroral occurrence frequency decreases at nights north of this zone

6.6 The Statistical Occurrence of Northern Lights

During the last part of the 19th century several physicists found a relationship between the occurrence of the northern lights and sunspots. In different countries scientists went back into the historical annals and looked for written statements of auroral observations. The most well known in this regard was the Swiss physicist and engineer Herman Fritz (1830–1893) who wrote his well-known book *Das Polarlicht* (Chap. 2.5) in 1881 in which he showed that the occurrence of the northern light has a maximum zone close to 67° north – called the auroral zone. He also studied the connection between the occurrence of sunspots and northern lights and found that a general relationship existed.

Lesser-known reports from this time concerning the same matter were made by people like Elias Loomis (1811–1889) in the United States, Rubenson (1829–1902) in Sweden and Tromholt (1851–1896) in

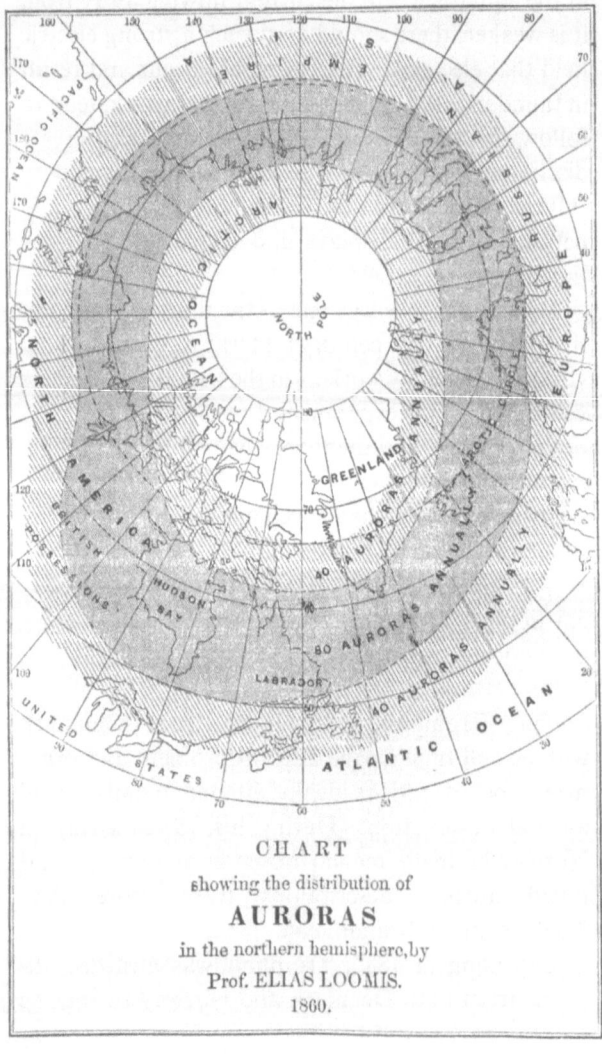

CHART
showing the distribution of
AURORAS
in the northern hemisphere, by
Prof. ELIAS LOOMIS.
1860.

Fig.6.12. Robert Rubenson (1829–1902) professor and director of the Swedish Meteorological Institute from 1873 collected historical auroral data and instructed his assistants to observe the northern light carefully from the network of meteorological stations in Sweden. He made a list of auroral data from the earliest date to about 1875.

Norway. Loomis derived a list of annual auroral occurrence numbers from about 1775 until 1873, and showed a period of very low auroral activity between 1795 and 1825 (see Fig. 10.3). He also made a map of the geographic distribution of the northern light and showed in his publication from 1860 that the maximum auroral frequency occurs in an oval zone around the globe passing the northern part of Norway, Iceland and the southern tip of Greenland. Loomis's work, which was published more than 20 years before Fritz's *Das Polarlicht*, has never achieved the same publicity as the latter although it essentially represents a similarly valid scientific result.

Rubenson, who was director of the Swedish Meteorological Institute, had a large collection of written auroral data from about 1720 to 1875. He used the network of meteorological stations in Sweden for auroral observations, probably the best organized network of his time. His results (see Fig. 10.3) are comparable to those of Loomis except that the period of low auroral activity at the beginning of the last century was shorter in Rubenson's data. Rubenson's data do not show the broad minimum of auroral occurrence between 1855 and 1870 as those of Loomis, did.

Tromholt collected data not only from Norway but also from other parts of Scandinavia and he made a list of observations starting in about 1720 and ending in about 1878. His data (see Fig. 10.3) are almost identical to those of Rubenson, but there are details that disagree, such as an outstanding maximum seen in 1839 (not present in Rubenson's data). This peak is in fact due to data from the French expedition to Alta during the winter of 1838–1839. Since Rubenson used only Swedish data, this peak is not present in his results.

This incident clearly shows how difficult it is to use auroral data based only on eye observations in order to deduce the true occurrence frequency of this phenomenon.

In later years it has been realized that the occurrence of a specially intense red type of northern light is very well correlated with maxima in the sunspot cycle. From the data by Tromholt this is indeed found to be the case (see Fig. 10.5). By going back to the old annals and looking for descriptions of such intense red auroras, one can in fact deduce information about the occurrence of solar activity.

6.7 Adam Paulsen Launched the Cosmic Theory

From a scientific viewpoint and particularly in physics, the last century was characterized by the solving of several long-standing problems and many new things being discovered. This was especially true in that branch of physics now known as electromagnetism. Attempts were made to incorporate these newly emerging physics disciplines and discoveries into auroral research. As a consequence, speculative theories, especially in informal gatherings in Europe, reached high popularity. Typical subjects for discussion at informal gatherings and parties during the previous century, were the auroral substance or the

Fig.6.13. Adam Fredrik Wivet Paulsen (1833–1907) a Danish professor in physics. He was appointed leader of the Danish expedition to Greenland in 1882–1883 in connection with The First International Polar Year. From 1884 he was the director of the Danish Meteorological Institute

material content of the northern light, meteor dust, iron bearing particles, etc.

One of those who most vividly made fun of these speculative auroral theories was Tromholt, and he wrote:

"A great deal of these theories, even some of the most recent, are so untenable that we have to explain their existence by the fact that they are made mainly by scientists from countries in the south, where the northern light is a rarity, and where it never occurs with the rich development that it achieves in its original home. One is surprised at this unusual sight, (the northern light), and after having observed a few, one immediately considers oneself to be capable of building up a theory about the whole phenomenon.

One can take almost the first and the best rated written material in a myriad of amusing books in natural sciences of our time where the northern light is discussed, and an expert on the subject will find one absurdity after another, as well as now and then there will be included a terrible picture, supposedly representing the northern light, but which does not belong anywhere other than in the illustrator's imagination. Without having first hand knowledge of the phenomenon one compiler discounts another, includes truth with untruth, the misunderstood with the distorted. The literature of the northern light is thereby, in the course of time, loaded with such a lot of traditional untruths that it would be a true Herculean task for somebody to ever clear up this Augean stable [14]".

The so called cosmic theory, which was a further development of de Mairan's theory, was one of those most discussed just before the turn of the century. However, this theory must take a back seat to Edlund's electric theory in popularity (Chap. 6.4).

When Tromholt and others had shown that there is a definite relationship between the occurrence of the northern light and the period of sunspots, the cosmic theory rode on a new crest of the wave. In the Nordic countries the leader was a Dane named Adam Frederik Wivet Paulsen (1833–1907). In an article called *The radiation theory of northern light* which was published in Nyt-Tidskrift for Fysik og Kemi in 1896, Paulsen argued against Edlund's induction theory. First of all he could not understand how the rotation of the earth could create currents in the upper atmosphere, and if it did he maintained that these currents would be counteracted by forces which would tend to move them in the opposite direction of the Earth's rotation. He therefore thought that Edlund's theory, if correct, would imply that all polar light had to move from east to west. He therefore proposed an alternative theory by the following words:

"We will then suppose that the Northern Light is a Radiation Phenomenon, caused by a Fluorescence in the Air due to Absorption of Rays comming from the uppermost Regions of the Atmosphere".

He continues:

"According to *The Radiation Theory* the Arcs and Bands do not themselves physically radiate Rays, but invisible Rays make the Arcs, and the Air fluoresces with enhanced Intensity also in the highest Regions, and by this Fluorescence the Patterns of the Rays are made visible".

Paulsen believed that the northern light could penetrate the atmosphere all the way down to an altitude of about 1 km; and that its uppermost part reached altitudes of 600 km above the ground. Due to this supposedly low altitude Paulsen believed that he could explain some auroral-like cloud formations. He found support for this theory in experiments with cathode rays in air. These rays produce ozone, a gas which causes the water vapour in the air to condense, and which in turn results in cloud formation. Finally Paulsen claims in excitement:

"I am now able to prove that the Northern Light is a Phenomenon created by Absorption of Cathode Rays [15] and am able to explain the Appearance of the Northern Light, its Relationship to the Earth's Magnetism and to the different Periods in the Intensity of the Phenomenon".

Paulsen also thought that the highest regions of the atmosphere were surrounded by a layer with an excess of negative electricity (electrons). The excess negative electricity layer was maintained by a steady influx from the Sun. The cathode rays would move from this layer towards the Earth, but because of the influence of the magnetic field, the rays could move further down into the atmosphere at the highest latitudes. Only in these latitudes could one therefore see structured forms of the northern light such as arcs and beams. At the lower latitudes, only a small, diffuse northern light could arise, because the magnetic field there prohibited the cathode rays from penetrating far enough into the atmosphere to produce striated northern lights. There were other properties of the northern light about which this theory could give no adequate explanation, for example diurnal variations and movements.

In spite of these apparently serious limitations, one could say that the cosmic theory was the most realistic of the theories of the 19th century, and this fact was probably then quite clear to most people.

6.8 Svante Arrhenius Expands de Mairan's Theory

The famous Swedish physicist and chemist Svante August Arrhenius (1859–1927) won the Nobel Prize

---

[14] Augean stable – filthy stable which Hercules cleansed by diverting the river Alpens through it
[15] Cathode rays are rays consisting of free electrons

Fig. 6.14. Svante August Arrhenius (1859–1927), a Swedish physicist and chemist. He was appointed professor in physics in 1895 and won the Nobel Prize in chemistry in 1903. In his later years he became interested in cosmic problems among which was the northern light

in chemistry in 1903 for his work on the electrolytic dissociation theory. In 1900 he published an auroral theory based on radiation pressure of electromagnetic waves and eruptions of particles from the Sun. He proposed that during great eruptions on the Sun, large quantities of matter, in the form of droplets or dust, are ejected in a radial direction from the sunspots and are pushed out from the Sun by radiation pressure (A mass transport which today is called the Solar Wind). He also said that during these eruptions on the Sun, negative and positive electricity are split into two and that this division produces cathode rays and X-rays. According to Arrhenius these rays have the ability to ionize the gas through which they propagate, and in this way negative electricity is created and the Sun is left positively charged. These outstreaming negative charges will hit other celestial bodies such as the planets, moon, etc., causing them to become negatively charged. They then repel the streaming negative charges from the Sun into hyperbolic paths away from the repelling body. The most rapidly streaming particles, however, are able to strike the body and increase its charge. The body cannot be infinitely charged and therefore has to be occasionally discharged. This discharge could also be triggered by ultraviolet radiation from the Sun according to Arrhenius's theory.

Arrhenius also believed that on the dayside of Earth at a height of 160–200 km there is an area in the atmosphere lying close to the central Sun-Earth line which would be strongly charged to a negative potential. By means of ultraviolet radiation from the sun, the discharges would primarily occur on the dayside in the equatorial region; the flow in the upper air could, however, move such charged particles to other time sectors of the Earth and thereby bring about discharges also on the nightside. Arrhenius claimed that since the discharges took place at a height of nearly 200 km the gas density is too low for noticeable light to be produced there. The cathode rays, however, must spiral along magnetic field lines and move toward the poles where they are forced downward into denser air and there form the northern light. No negative charges are able to reach the poles, however, and for this reason an auroral ring will be formed around each of the poles according to Arrhenius.

With this theory Arrhenius could explain the high correlation existing between the occurrence of the northern light and sunspot activity. On the basis of this theory, occurrence of the northern light would depend on the seasons since distance from the Sun would strongly influence the Sun's radiation pressure; furthermore Arrhenius's theory implies that the northern light will occur more frequently on the dayside than on the nightside. He in fact suggested that the occurrence of the northern light would be most frequent when the Sun is in the plane of the magnetic declination on the dayside.

With this theory Arrhenius was able to modernize de Mairan's work sufficiently to link together more than 200 years of speculation and the modern theory of the solar wind.

At the turn of the century the famous Danish polar explorer Peter Elfred Freuchen (1886–1957), who assisted Adam Paulsen on his expedition to Greenland, made a couple of reviews of the stages of auroral physics and concluded that:

"One has naturally been concerned about the visible northern lights, that is those which show up during the night or in the winter darkness, but it is possible that the daylight hides a large amount of northern lights which ought to be taken into account and which might be detected spectroscopically or by magnetic disturbances. The thought is near that the auroral gleam is always present in the sky in the northern areas; this may be the reason for the northerly winter nights not being as dark as one would think".

6.9 The Concept of the Auroral Ring

When F. C. Mayer (Chap. 7.3) in St. Petersburg at the first half of the 18th century introduced his methods for triangulation of the auroral height he assumed that the auroral arcs formed a complete ring around the polar cap and were centred at the geographic North Pole. Carl Frederik Fearnley (1818–1890) a Norwegian astronomer, However, when adopting Mayer's methods, assumed that the centre of the auroral ring was at the magnetic North Pole, a concept he used in order to triangulate the auroral altitude by observations of auroral arcs from one point only.

Adolf Erik Nordenskiöld (1832–1901) who was in charge of the famous "Vega" expedition and was the first ever to travel through the North-East passage in 1878–79, assumed for his altitude triangulation work that the center of the auroral ring was somewhere between the geomagnetic and geographic north poles. In his reports from the Vega expedition he devoted a large section to the northern light and depicted the auroral ring (Fig. 6.16) with its centre very close to the so-called geomagnetic north pole. Nordenskiöld called this point "the pole of the northern light".

When Weyprecht (Chap. 3.4) who initiated the First International Polar Year (1882–83) and later

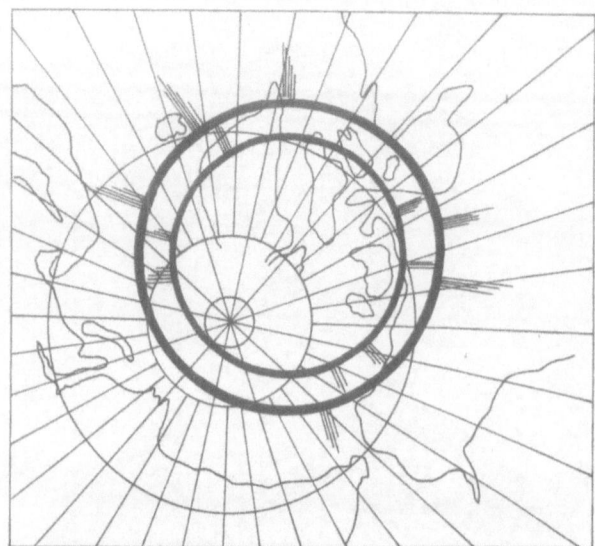

Fig. 6.16. Nordenskiölds concept of the auroral ring with its centre at the so-called pole of the northern light. Today it is clear that this point is very close to the geomagnetic North Pole. This figure is redrawn from Nordenskiölds Om Norrskenen, Stockholm (1881)

Tromholt (Chap. 6.5) discussed the motion of the auroral ring with respect to the solar sunspot activity, they revealed an insight into the morphology of the northern light which was much provident.

The concept of an homogeneous illuminated auroral ring around the polar cap was little appreciated during the first half of this century. Not until the International Geophysical Year (1957–58) was a great effort put forward to study the global distribution of the northern light. From this enterprise it was clear that the northern light at any moment was distributed in an oval belt around the polar cap, and this belt has since been named "the auroral oval". Its centre was found to be very close to the geomagnetic pole. It was also found that the size of the oval expanded as it moved to lower latitudes when the auroral activity increased.

This oval was not homogeneously illuminated as Mayer and the others believed, but was broken up into segments of varying auroral forms and strength of illumination. After years of observations by satellites from space, it is now getting clear that the auroral oval is much like a circular zone as Nordenskiöld and Tromholt depicted it and that its centre is very close to the geomagnetic pole or Nordenskiöld's pole of the Northern light.

Fig. 6.15. Adolf Erik Nordenskiöld (1832–1901), Finnish of birth, who later became the leader of the Swedish "Vega" expedition which sailed through the North-East Passage as the first vessel ever

7 Norwegian Auroral Pioneers in the Dawn of Our Century

7.1 Professor Carl Frederik Mülertz Størmer Mapped the Auroral Forms

Towards the close of the 19th century there began a rapid increase in scientific discoveries in many different fields and included among these were the contributions from Norwegian auroral physicists. At this time knowledge about the northern light had been obtained to an overwhelming extent from visual observations. In order to better understand the problems facing these auroral physicists, we digress momentarily to discuss the visual aurora.

The aurora can vary in visibility from a barely detectable glow among the stars to a luminance bright enough to completely veil the stars at the place where it occurs. Auroral emissions also occur at wavelengths where the eye is insensitive (e.g., in the ultraviolet and infrared portions of the spectrum) and at luminances below the threshold level of the eye. The eye is not equally sensitive to all wavelengths of light, and in addition, at a given wavelength its sensitivity depends on its state of dark adaptation. In full daylight, it is most sensitive to radiation at 5,600 Å, whereas when fully dark adapted the peak sensitivity occurs at 5,100 Å. The eye is sensitive to light from 4,000–7,000 Å. [16]

---

[16] 1 Å = 1/10,000,000,000 m

The unit of brightness measurement used in auroral research is the Rayleigh (R) which is based on the flux density of photons – this unit does not take into account the sensitivity curve of the eye. For this reason an aurora of intensity approximately 0.7 to 1 kR (1 kR = 1,000 R) at 5,577 Å (the green line) becomes faintly visible whereas 5–7 kR at 6,300 Å (the red line) is necessary in order to be fully visible. The brightness at which a northern light reveals its colour occurs at a somewhat higher value than the threshold level for detection.

The table below shows how auroral intensities compare with other celestial objects.

Table 7.1

Light source	Light flux in $mW\,m^{-2}$	Number of photons cm^{-2} (column) and s^{-1}
The Sun	1.4×10^6	
Full Moon	3	
Very strong northern light[a]	1	$10^{12} = 1,000$ kR
Strong northern light	1/10	$10^{11} = 100$ kR
Medium strong northern light	1/100	$10^{10} = 10$ kR
Faint northern light	1/1,000	$10^9 = 1$ kR
Invisible northern light	1/10,000	$< 10^8 < 0.1$ kR
Total starlight	2/1,000	
Night sky light[b]	2/100	
Cosmic light	4/1,000	

[a] The intensity of the northern light can vary over more than four orders of magnitude. At maximum intensity the illumination on the ground can be comparable with moonlight, and it is ~100 times stronger than starlight

[b] The night skylight comes mainly from the air at heights of 50–100 km. The gas here is warmed up by the Sun during the day and cooled down during the night. While cooling down the gas gives up some energy in the form of weak light (W = Watt, mW = 0.001 W)

Fig. 7.1. Carl Fredrik Mülertz Størmer (1874–1957) was appointed professor in pure mathematics in 1903 at the University of Oslo. Professor Birkeland introduced him to the problems in auroral physics in 1902, a subject to which Størmer devoted most of his life

Fig. 7.2. A drawing of professor Størmer made by the artist Øyvind Sørensen in Størmer's office at the University of Oslo. Here the heights of the northern light were derived from the 40,000 pictures taken in the field. For this analysis he used two projectors which are shown on the drawing

During his long lifetime Professor Carl Frederik Mülertz Størmer (1874–1957) devoted most of his time and energy to solving the riddle of the northern light (Fig. 7.1). He began an accurate, quantitative auroral study and carried it out in an excellent manner. His theoretical calculations also became far more important and wide ranging than his application of them to the northern light.

Over a period of approximately 40 years Størmer and his co-workers took more than 40,000 pictures of the northern light, and in about half of these the height of this phenomenon was measured. Størmer's measurements are still the basic ones for our knowledge of the height distribution of the different northern lights.

During an intense auroral display one gets the impression that it consists of a confusingly large number of different forms and structures. Størmer studied these in detail, defined them and described characteristic features of the different forms which altogether comprise an auroral atlas.

Until about 1950 it was customary to classify the northern light into about 15 different forms. With simultaneous ground and satellite observations it is now possible to catalogue all observations by use of the following four primary forms:

a) Homogeneous quiet arcs and bands. Auroral arcs and bands (see Fig. 7.3) are by far the most prevalent. The light stretches along arcs and bands across the sky in an approximately east–west direction. The auroral bands differ from the arcs in that they are more irregular in form and extension. The width of the arcs and bands may vary from 1–100 km, while their lengths may often exceed 1,000 km. Different parallel forms frequently appear simultaneously. Examples of arcs and bands are shown in Fig. 7.3, and may be seen in the photographs on pages 45 to 51.

b) Active northern light with ray structure. The northern light may be split up into long, thin (nearly

vertical) rays and these lie along the magnetic field lines. The length of the rays may vary from some tens to many hundreds of kilometers. Auroral rays sometimes occur as an independent structure forming a corona of draperies, or may changes shapes into arcs and bands with rayed structure. Northern light with rayed structures can be seen for example in Fig. 7.3 and on pages 24 and 45 to 51.

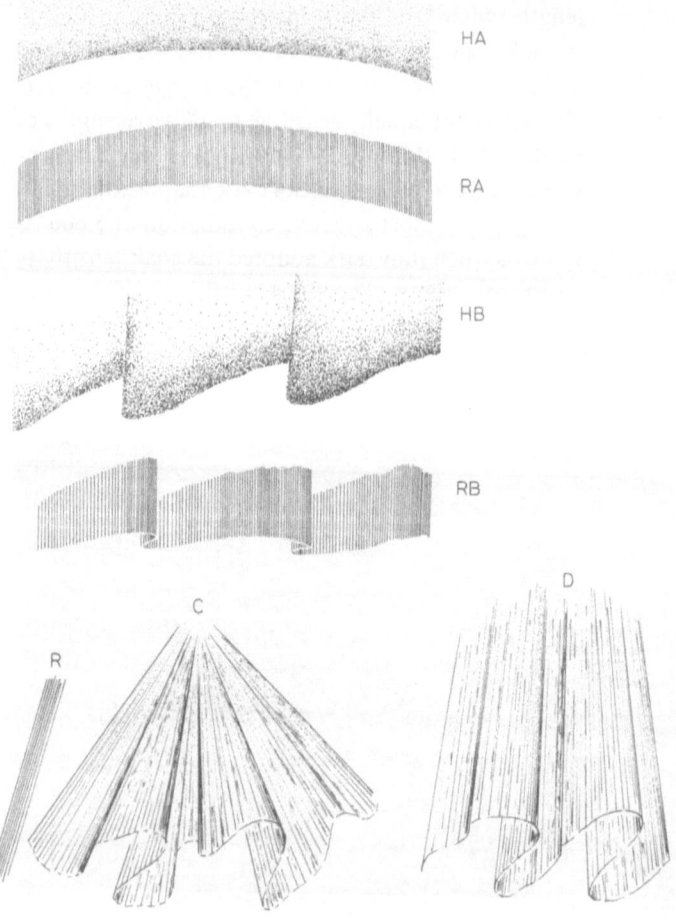

Fig. 7.3. Illustrations of typical forms of the northern light. *HA* homogeneous arc; *RA* arc with ray structure; *HB* homogeneous band; *RB* band with ray structure. The three lower forms are rays (*R*), corona (*C*), and draperies (*D*)

88

c) Diffuse spots and surfaces. Other typical forms are diffuse, cloudlike surfaces which are usually greyish-green in colour. These surfaces often cover several hundred square meters in area and occur mainly in the morning before dawn. Because of their diffuse character – i.e. having little structure – they are often difficult to see with the naked eye.

d) Spiral structure. When there are intense auroral displays several different spiral structured forms occur. Typical dimensions of these structures are 10–100 km. The spirals are often curled in different directions (see colour photographs on pp. 45–51).

Homogeneous arcs and bands are classified as quiet forms, while rays, coronas, draperies and spirals are more active, mobile forms (cf. Figs. 3.1, 3.3, 3.5 and 7.3).

Early in the evening or some hours after midnight one often sees northern light lying almost totally quiet or moving very slowly, i.e., a few tens of meters per second, while much greater speeds, e.g., up to 100 km per second are often seen during great auroral displays. During these times auroral forms move about in the sky and, in particular, small waves of light move about from one form to another at an incredible speed.

In particular, on the morning side diffuse auroral surfaces and spots appear, which are difficult to see with the naked eye. These light veils can be stationary for hours. They become bright and decay at intervals of a few seconds, without the forms radically changing in shape. This is what is termed pulsating aurora, and the phenomenon resembles puffs from a locomotive. The pulsations sometimes are in phase or almost in phase throughout the entire form. At other times one can see flamelike light waves spreading periodically up the sky with great speed.

7.2 Størmer Mapped the Geographical Distribution of the Northern Light

Our ancestors have known for several centuries that the northern light mainly appears in the north sky; in Scandinavia, however, it often appeared in the zenith. The geographic distribution of the northern light was one of the first problems to be given a more systematical analysis.

Until the 1850's it was believed that the frequency occurrence of the northern light increased with increasing latitude from the equator and that they occurred most frequently at the poles. Therefore it was a great surprise to many when the American physicist Elias Loomis (1860) and the Swiss engineer Herman Fritz (1881) proved that auroral activity and auroral intensity were greatest in an approximate 500 km broad, circularly shaped zone about 2,000 km from the magnetic pole (Fig. 7.4). The geographic distribution of the auroral zone in Norway is along the coast of Troms and Finnmark. The idea that the northern light mainly appears in two narrow, circular zones around the poles was mentioned in a work by the German geographer Muncke as early as 1833 and by Wargentin and Jessen in the 18th century (Chaps. 5.9 and 5.14).

Størmer made a thorough investigation of the northern light. He found that auroral occurrence frequency diminishes rapidly with distance from the auroral zone. The light can be seen on practically every clear night in north Norway whereas in the Oslo area it can only be seen, on the average, three or four times per month. If one goes 10° in latitude south of Oslo one can probably see the northern light on a clear night once a year. Also, as one travels farther toward the equator, the polar light becomes less colourful and inactive. In the Mediterranean area, for example, the northern light will be seen only a few times in a hundred years, and when this does happen it becomes a very big news event.

The polar light occurs with the same frequency in both the southern and northern hemispheres of the Earth. It is now well understood that the polar light occurs in both zones simultaneously and they are almost exact mirror images of each other (see the colour photograph on p. 121). But because of the dissimilar light circumstances (i.e., winter in one hemisphere at the same time as summer is in the other) there are only short periods of time when observations of the conjugate polar lights can be made simultaneously in the north and south auroral zones.

The northern light also occurs poleward of the auroral zones but its intensity and motion is markedly less. The positions of the auroral zones can change depending upon solar activity. During times of great solar activity the zone will move more equatorward than during quiet solar conditions.

Norway is located in an especially favourable position for auroral observations. It has a long geographical extent (north–south) where North Norway is situated in the auroral zone. This ideal geographic location is probably the primary reason that Norwegians have made pioneering contributions in auroral physics.

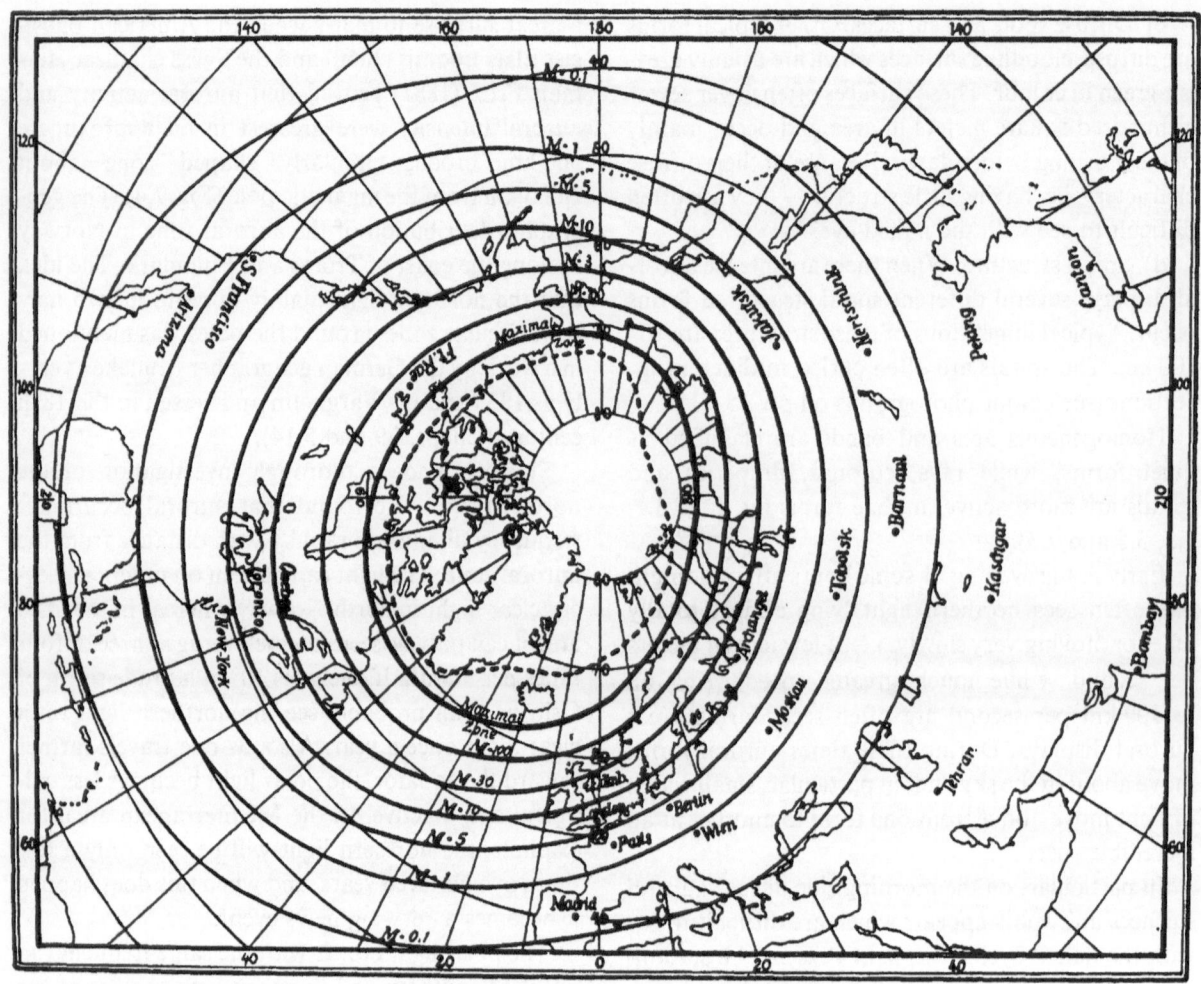

Fig. 7.4. Fritz's map of the so-called isochasms, i.e., curves showing places where the occurrence frequency of the northern light is the same at midnight. The maximum zone marked here encircles the geomagnetic pole (marked with a *small ring*) and passes outside the coast of Northern Norway, the northern tip of Novalja Zemlja, south of Iceland and Greenland. The *dotted line* marks the ice barrier, and the magnetic pole is marked by a *cross*. This figure should be compared by the one made by Loomis (Fig. 6.11)

7.3 Størmer Accurately Determined the Height of the Northern Light Once and For All

a) The Height of the Northern Light – the Most Disputed Question. Of all the characteristic properties of the northern light until Størmer's time, none had been studied and discussed more than its height. In the approximate time period between the 18th and the beginning of the 20th century it was a controversial question which was frequently debated. Knowledge about the height and geographic location of the northern light was necessary in order to

understand and formulate a realistic auroral theory. At the same time a well-determined height of the northern light would yield vital information about properties of the atmosphere at much greater distances than previous, direct or indirect, ground measurements had been able to accomplish.

All reported auroral heights before 1910 were based exclusively on visual observations and consequently there were large inaccuracies in the measurements.

The first known method for measuring auroral heights goes back to about the year 1730. Scientists at this time thought that the Earth's stratum of air was very thin (10–20 km) and beyond this stratum there was a vacuum or aether. It was believed that the northern light occurred in this thin air layer.

Dr. F.C. Mayer who did his work in St. Petersburg (now Leningrad), Russia, is credited with developing a method for height measurement of the northern light. His method assumed that the northern light occurred in regular circular auroral segments, and in

Table 7.2. Measured heights of the northern lights

Observers	Sites	Time period	Number of observations	Minimum height (km)	Maximum height (km)	Average height (km)
Gassendi	Peinier	Sept. 1621	1	–	–	850
Kraft	St. Petersburg	Autumn 1731	3	200	680	370
Cramer et al.	Geneva	1730	2	660	770	–
Horrebow	Copenhagen	1731–1736	5	680	1,030	850
de Mairan et al.	Paris	1731–1751	5	650	1,000	770
Bergman et al.	Uppsala	1759–1764	11	380	1,300	760
Celsius	Uppsala	1740	2	650	800	–

practice this method was limited to stable auroral arcs. From observation one determined the arc's position in the sky (i.e., the distance in degrees from the vertical) and its length in degrees. With this information the height can be calculated.

About the same time that Mayer did this work, the Frenchman de Mairan worked out a method for height measurement based on simultaneous observations from two stations spaced 10–20 km apart. If the distance between the positions of the two observers is known, the height can be found. In practice, the main problem is for the two measurements to be made simultaneously on the same point in the northern light. This method is called "the parallax method" (Fig. 7.5).

Both Mayer's and de Mairan's methods require that a quiet auroral form be used in the measurement and therefore practically all height measurements were attempted on quiet auroral arcs. In 1764,

T. Bergman (Chap. 5.15) made a comparison of height measurements which were known at that time. The important parts of this work are given in Table 7.2. All of these auroral heights as measured were somewhat greater than 100 km. On the other hand, several "travel reports" existed where it was concluded that the northern light most ofen occurred near the ground and at some distance below the clouds.

In the following is an example of one such report dating back to January 1882:

"We had just reached to the middle of Korsfjorden when I suddenly discovered a northern light above Alta. It had turned itself into a bundle close to the water surface and came rushing with an incredible speed towards us out along the fjord. I discovered at once that it was aiming at the boat, and screamed to my rowing friends which were sitting with their backs turned against the northern light. "It throws the boat over", Jakob screamed from the main thwart. I could see through the dark that the men were bending their backs and rowing like crazy by

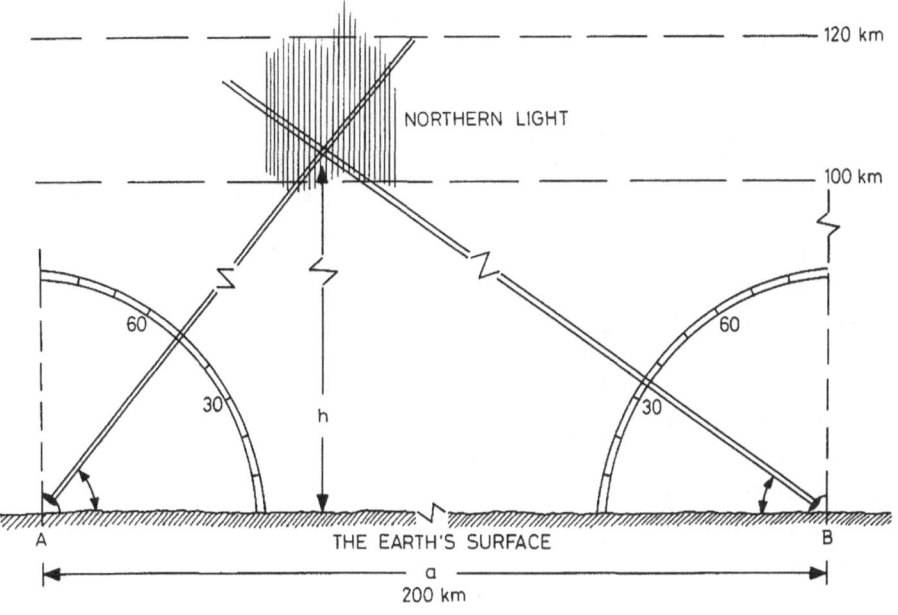

Fig. 7.5. An illustration showing the parallax method for measuring the height of the northern light. Two observers, one at A and one at B at a known distance from each other, read the angular distance to the same point in the northern light. It is then relatively easy to derive the height, *h*, of the light display when the curvature of the Earth is taken into account. The main problem related to this method is to ensure that the observers actually are aiming at the same point in the display

91

striking the oars in the sea so that the phosphorescence were rolling and lighting against the oarblades. It was necessary to keep the boat up against the furious weather which now had reached very closely towards us. I was also aware that if a puff of wind was following the accelerating northern light, the boat would be swallowed as by a gigantic waterspout. It was something quite different from a real storm. I had never heard before by the elder people that the northern light ever could go astray down to the sea level to play boogeyman in the fjords ..."

b) Auroral Height Measurements in the Twentieth Century. During the 19 th century not a great deal of progress was made in attempts to determine the height of the northern light. We shall mention here some of the results from the many active scientists who published more or less systematic observations.

The famous English natural scientist Dalton reported auroral heights of about 150 km based on observations made around 1820. His colleague Potter, around the year 1835, thought that the average height should be about 100 km. The French expedition which set up winter quarters in Bossekop (Alta region of north Norway) during 1838–1839 carried out measurements using the parallax method and came to the conclusion that the auroral heights varied from 90–150 km.

Around 1850 the Norwegian astronomer Carl Frederik Fearnley (1818–1890) performed some height measurements in Kristiania. He arrived at a mean height value of 200 km based on 16 observations. In the same time period 20 measurements were made at Greenland where the height measurements varied greatly from 600 m to 68 km. Among the better-known auroral scientists of the previous century, Fritz and Lemstrøm (Chaps. 6.6 and 6.3) both thought that the northern light could occur all the way down to the Earth's surface, while Loomis reported heights from 70 to 870 km.

The First International Polar Year in 1882–1883 was of great importance for auroral research but determining the aurora's height accurately remained a big problem. The Norwegian Polar Expedition in 1882–1883 concentrated on auroral observations in Finnmark from Kautokeino and Bossekop. There were height measurements with values ranging from 76–163 km which gave a mean height of 113 km.

c) Størmer's Measurements of the Heights of the Northern Light. The precise, quantitative and thorough auroral studies made by Størmer exemplify the method which he used in his auroral investigations. The manner in which he charted auroral heights and positions in the atmosphere is described by him as follows:

"The only reliable and objective method (to determine the height of the northern light) is photography, but for a long time all attempts at photographing the northern light were in vain. While on a trip to Bossekop in 1892, two German scientists by chance took a photograph of an intense auroral drapery with an exposure time of only seven seconds. To my knowledge this was the only picture taken with such a short exposure time up until 1909 when I began some systematic experiments to photograph the aurora. By testing and comparing, I was able to find a small, extremely "fast" camera lens which gave excellent results. Using this lens and the most violet-sensitive photographic plate available, I succeeded in taking pictures of a bright northern light with an exposure time of only 1 s or less. When the problem of auroral photography was satisfactorily solved, I went on two expeditions in 1910 and 1913 to Bossekop to photograph the northern light and determine its height simultaneously from two stations separated by an appropriate distance – the two stations also being in contact with each other by telephone. On the photographic plates, star patterns were also recorded and from the different positions among the stars as seen from the two stations, the height could be calculated from measurements on the photographic plates".

The photographic plot gives an almost true representation of the actual picture as compared to the eye with which one sees or thinks one sees. The height measuring problem was tackled by Størmer with a strictly scientific discipline. He employed a rational procedure which in principle allowed a reconstruction of every single configuration in three dimensions. Størmer accurately measured the baseline between the two stations and provided each with a telephone. He was the pioneer who introduced a methodical and logical scientific method into auroral height measurements. Some of his results are shown in Fig. 7.6 and Table 7.3.

Table 7.3. Average lower heights of northern lights measured in Norway by Størmer and others

Auroral forms	South Norway		North Norway	
	Average height	Number of observations	Average height	Number of observations
Homogenous arcs	106	161	109	355
Arcs with ray structure	101	32	107	888
Homogenous bands	94	10	–	–
Pulsating arcs	103	434	–	–
Pulsating surfaces	93	19	106	160
Diffuse surfaces	93	204	–	–
Rays	127	52	113	61
Bands with ray structure	90	531	110	409
Draperies	104	64	–	–

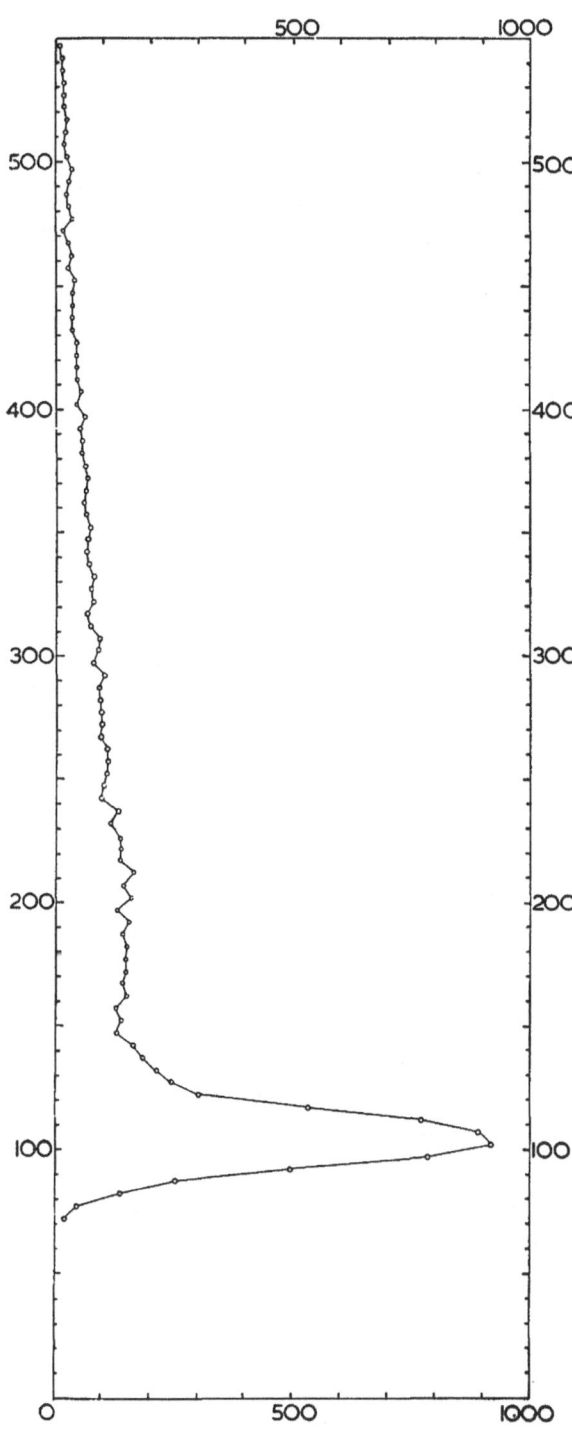

Fig. 7.6. The statistical distribution of the more than 10,000 height measurements of the northern light derived by Størmer and his colleagues. The vertical scale gives the height in kilometers while the horizontal scale gives the number of measurements. Most of the northern lights are found between 90 and 150 km

Bygdø (near the waterfront in Oslo) and was allowed to use the old observatory in Kristiania (now Oslo). In addition to these two stations, Størmer established others at suitable distances from Kristiania: Kongsberg, Lillehammer, at Tomte in Hurdal, Dombås, Askim, etc. These stations were kept in operation by different people until 1957.

In a newspaper article in 1927, Størmer described his experimental method as follows:

"The principle of auroral height measurements by photography was quite simply the following: We establish two auroral stations in contact with each other by telephone so that I can at any time discuss the auroral situation with my assistants. The distance between the stations was about 30 km. After a telephone conversation we adjusted the photographic apparatus such that the same stars, as nearly as possible, would be in the field of view of the northern light of interest. The exposure occurs after the signal is made and we simultaneously take two photographs of the same display – one from the main station and the other from the auxiliary station. We see that the northern light lies at different places in the star background on the two plates, and from this difference the height and position of the northern light can be computed when all the necessary data from the plates is compared and calculated. The necessary calculations are tedious and the raw data recorded in one evening's observations often requires months of processing".

From an analysis of some photographs in 1926 Størmer discovered the so-called "solar illuminated aurora". This is a type of northern light which is observed only during special twilight conditions. At such conditions the Sun can illuminate the extreme upper parts of the atmosphere while the lower parts (e.g., up to 200 km) are still in darkness.

The solar illuminated auroral rays generally reach greater heights than those which are situated in the darkness. The realization that the northern light could occur as high as 1,000 km stimulated exploration of the outermost parts of the atmosphere and ionosphere to a considerable degree. All of the atmospheric models dealing with composition and density at an altitude around 1,000 km had to be completely revised. Therefore, after 1926, observations of sunlit auroral rays became a basic part of Størmer's research program.

Størmer's method was used extensively by other groups, and is still applied in auroral height measurements. Table 7.4 illustrates some of the measure-

The northern light is something which varies greatly in time and geographic position. In order to map out the height and position of the northern light it was necessary to monitor the displays continually by an accurate method.

Størmer began this task in his customary energetic manner and drew up his plan of attack in record time. He established one station in his home at

Table 7.4. Number of parallactic height measurements of northern lights

Sites	Time period	Observers	Number of observations analysed
South-Norway	1910–1943	Størmer et al.	13,500
North-Norway	1929–1936	Harang et al.	3,000
Greenland	1938–1939	Størmer et al.	500
Alaska	1930–1934	Fuller and Bramhall	1,600

ments made using his technique before the Second World War.

The way in which auroral heights vary for different northern light displays is given in Table 7.3. It can be seen that arcs and bands occur at the lowest heights while auroral rays lie at the greatest distance from the Earth's surface.

7.4 Professor Lars Vegard Studied the Colours of the Northern Light

a) The Colour Composition of the Northern Light is the Atmosphere's Fingerprint. In order to understand how a gas emits light, it is necessary to discuss the basic building blocks of material substances – i.e., the atoms and molecules. The atom is composed of a central, positively charged nucleus with one or more lighter negatively charged electrons circling at different distances around it. (The electron is the smallest carrier of electricity.) When solar particles enter the atmosphere they collide with atoms and molecules in the atmosphere and are slowed down. The northern light is a direct result of such collision processes. What physically happens when solar particles hit the electrons of an atom is that the courses of these particles are changed. The following two things can happen: (1) The electron can be freed from the nucleus – the gas is by this process said to be "ionized", or (2) the electron begins to circle in the area of the nucleus in a new orbit – in this case the gas is "excited". The excitation process merely moves the electron from one orbit to another and the whole particle has an excess of energy.

Fig. 7.7. Lars Vegard (1880–1963) a Norwegian professor in physics standing beside his spectrograph. Vegard was Birkeland's assistant from 1905 and was appointed professor in physics at the University of Christiania (now Oslo) in 1918. Vegard was one of the modern pioneers within auroral research and he contributed strongly to the understanding of the composition of colours in the northern light

An excited particle is very unstable. In one millionth of a second (10^{-6} s) it will give up this excess energy which it acquired in the excitation process. It is possible for the electron to return to its original shell – the stable ground state – and the excess energy is emitted in the form of a small twinkle of light or photon. In order to produce a visible northern light display, there must be several hundred of these photons emitted per cubic centimeter per second.

At auroral altitudes, the atmosphere is composed primarily of oxygen and nitrogen in both atomic and molecular form.

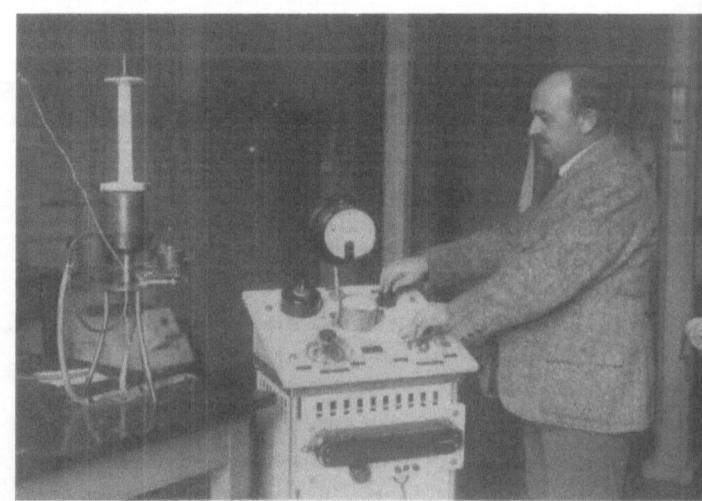

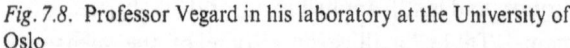

Fig. 7.8. Professor Vegard in his laboratory at the University of Oslo

The auroral colour spectrum is not continuous but is composed of a series of spectral lines and bands. Photons which produce the northern light must therefore originate in a gas since a solid material emits a continuum of light. The colour of each photon is determined by the gas particle from which it comes and its particular excited state. By measuring the wavelength of the light rays emitted by known gases in the laboratory, one can identify different gases in a mixture. By studying the colour composition of the northern light one can determine the constituents of the gases from which the northern light originates as well as the gases composing the Earth's atmosphere. By this means it can be seen that the northern light is the atmosphere's fingerprint.

In addition, the relative intensity of the different lines gives information about the abundance of the different gas particles in the atmosphere. For example, if the density of one type of gas particle is low, then we would expect weak light emission from this gas.

b) Vegard's Research on the Colour of the Northern Light. Prior to 1910, the only auroral colour with a fairly accurately known wavelength was the green line. This was the existing situation when Professor Lars Vegard (1880–1963) began his pioneering work on the auroral spectrum in the first half of the 20th century. Vegard measured the wavelength of the auroral green line with much more accuracy than Ångstrøm had done and obtained a value for it of $5,577 \pm 0.1$ Å.

Before 1920, the main problem was that no one had been able to demonstrate that this green auroral line was emitted by any of the known atmospheric gases. The auroral green line, therefore, did not reveal any secrets, but instead remained, more and more puzzling. This mystery brought about many bitter disputes and deterred progress, to a great extent, into investigations of the northern light. Several new gases were suggested as being responsible for the green line. Bold, but well considered hypotheses were proposed. We shall not go into any more details here on this matter other than to disclose how Vegard "got his feet wet" in the process of clearing up the green line mystery.

Vegard's hypothesis was that at high altitudes, the temperature was so low that nitrogen gas existed in a solid state – as ice crystals – in the atmosphere. In the mid 1920's, he performed some experiments in the laboratory and discovered that nitrogen in its crystalline form emits a green line which is very close in wavelength to the auroral green line. This made big news in the professional journals, and in the mass media the headlines read: "The Auroral Riddle Solved".

At about the same time, two Canadian scientists found the correct solution to the problem – namely that the auroral green line is emitted by the oxygen atom, but it is a "forbidden" transition which was very difficult to simulate in the laboratory at that time. This is because the oxygen atom can remain in its excited state over a time interval close to one second; i.e., the auroral green line is not emitted until about a second after the oxygen atom has collided with the electron.

Vegard was an active observer. After only a few years of measurements, he had accurately compiled about 40 different auroral colours. He made new discoveries literally one after the other, beginning in 1910. Vegard found that the blue-violet colour in the primary spectrum is due to an emission from molecular nitrogen. The red lower border of the northern light is mainly due to nitrogen gases, while the upper

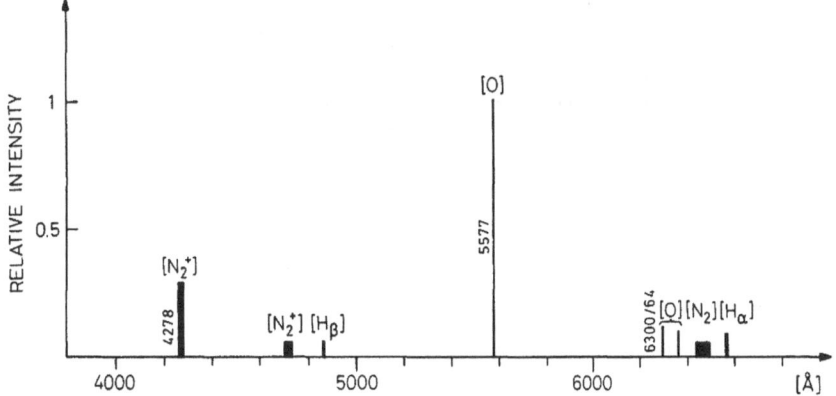

Fig. 7.9. The illustration shows the wavelength in Ångstrøm ($1 \text{ Å} = 10^{-10}$ m) and the relative intensity of some of the most prominent light emissions in the visible northern light. The green line at 5,577 Å is given an intensity of unity and is by far the strongest; the molecular band of N_2^+ centered at 4,278 Å is also a relatively strong emission. *O* represents oxygen atom, N_2 nitrogen molecules and *H* hydrogen. A colour picture of the auroral spectrum compared with a spectrum of the Sun is shown on p. 122

red colour is due to atomic oxygen. The auroral spectrum also contains a number of other emissions both in the visible part of the spectrum and in the ultraviolet and infrared (Fig. 7.9). Light from the minor atmospheric gases, such as hydrogen and helium, were not found. This was surprising because at this time it was believed that these were the dominant existing gases at auroral altitudes. Studies of auroral colour, therefore, were stalled until more realistic model atmospheres of composition and density could be determined in the height interval from 80 to about 1,000 km.

c) *Auroral Colour – a Means of Measuring the Particle's Speed.* The very important discovery which made Vegard most famous was that of the proton aurora. He strongly argued that since auroral particles are electrons which come from the Sun they must be accompanied by a positive atomic nucleus, preferentially protons. If this were not true, a polarized electric field of sufficient strength would be established, between the electron clouds and the Sun, for movement of the electrons to be stopped. When these Solar Wind protons bombard the atmosphere, he thought they must also produce a northern light. In 1952, Vegard published a popular article in which he wrote:

"Previously in 1939–1941, it happened that direct evidence came to me that these pencils of rays are produced just as I assumed (by electrons and protons). In 1939 in Oslo we recorded a series of auroral spectrogram where hydrogen spectral lines appeared with considerable strength. This showed that there are times when the Sun sends out positive ions (protons).

In the following 2 years we took several spectrograms of strong hydrogen lines. At The Auroral Observatory in Tromsø during both winters of 1939–1940, we made a series of strong spectrographic exposures, with a spectrograph having considerably better dispersion than the one we had used in Oslo. With the aid of this instrument, one could establish for the first time with absolute certainty that we were dealing with the hydrogen-line and in addition that on some of these spectrograms the hydrogen line had spread out into the form of a small band and had also strongly shifted toward shorter wavelengths. This broadening and shifting toward shorter wavelengths could not be attributed to defects in the picture because other nearby lines were very sharp. Therefore, the only explanation was that the hydrogen atoms, which produced the light, were moving at high speed and preferentially in the direction toward the observer; or at any rate down toward the Earth".

This discovery by Vegard opened a new window into space. In order to understand its importance, we must describe and illustrate the Doppler Effect. Let us imagine an ambulance is sounding its siren while passing by. We hear a tone from the siren which changes in pitch; from a higher one when it is approaching to a lower one when it is receding. We know that an automobile horn emits a constant tone, but as the automobile approaches, the sound waves from the horn are pressed closer together than if the horn was stationary, and we hear proportionately more sound waves per unit time. When the automobile is receding, we hear comparatively fewer sound waves per unit time than before, and it seems as though the wavelengths are stretched out, which makes the pitch sound lower. This change of pitch due to movement of the source relative to the observer is known as the Doppler Effect. This illustrative example used sound waves, but light waves act in the same manner for moving sources with respect to an observer and also produce the Doppler Effect (Fig. 7.10).

It was this same effect which Vegard observed in the proton aurora. When positive protons penetrate the atmosphere, they produce the hydrogen line in the northern light. If the proton is moving toward the observer, the wavelength of the observed hydrogen line decreases; that is, it is shifted towards the blue. The auroral hydrogen line functions,

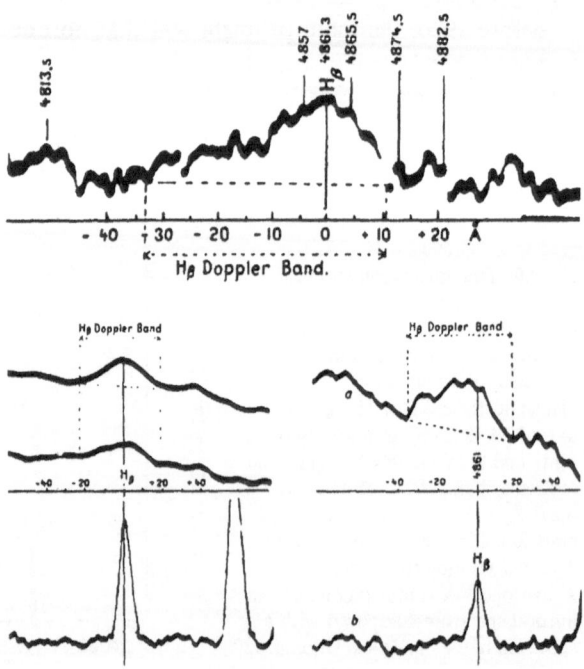

Fig. 7.10. Some of the original observations made by Vegard showing the Doppler broadening of the H$_\beta$ line in the proton aurora. The H$_\beta$ line without Doppler broadening and Doppler shift is shown at the bottom for comparison

96

therefore, as a kind of speedometer for the incoming proton. The amount by which the line is shifted is a measure of the speed of the bombarding particle (the proton). The more the proton light is shifted towards the blue – the greater is the speed of the proton which produced it.

Therefore, the conclusion is that Vegard's demonstration of the existence of the proton aurora was "dotting the i" or "putting the cream on the cake" in regard to Birkeland's auroral theory (Chap. 7.5). Everyone was in agreement that the northern light was due to a collision between the Solar Wind particles and atmospheric gases, but Vegard's measurements showed this directly and also gave indirect information on the speed of the particle.

Vegard's observations also showed that protons bombard the upper atmosphere and produce northern lights. These weak hydrogen lines – the first ones in the Balmer series – are called Hα (6,563 Å) and Hβ (4,861 Å). They are created by energetic protons which bombard the atmosphere (the proton is a positively charged ion – or a hydrogen atom which has lost its electron through the ionization process (Chap. 7.4a). The proton captures an electron and forms a hydrogen atom.

In the capturing process, the hydrogen atom can find itself in a higher than normal energy state – that is, in an excited state. In the process of returning to its normal ground energy state, it will emit photons (the so-called "proton aurora"). Following this event, the hydrogen atom may collide with another particle and once again create another proton such that the process can start all over again and continue until the energy of the incoming proton is exhausted.

This effect also reduces the grip the magnetic field has on the penetrating protons since the excess energy can be shifted over to the hydrogen atoms which can move freely about in the atmosphere. Therefore a typical proton aurora will be much more smeared out than a typical electron aurora. In the latter case the penetrating electrons are forced to move closely along the magnetic field lines, and therefore the electron aurora has much more structure like arcs, bands, rays etc.

d) Auroral Colours as Related to Temperature. Gas temperature is an important parameter which can be determined from auroral studies, and it was Vegard who pointed out that auroral emissions contain information about the gas from which the light originates. The auroral emission colours will change with temperature variations in the gas. The temperature of the gas can be determined from the light emitted by either of two means:

1. Accurate measurements of the width of the auroral line, or
2. The intensity ratio between the different emission bands of molecular gases.

In the upper atmosphere only a small fraction of the total number of atoms and molecules emit light. Strictly speaking it is temperature information about this part of the atmosphere alone which we obtain from auroral measurements. In using auroral observations for this purpose, one must assume that the atoms and molecules which emit the light are in temperature equilibrium with the gases.

The molecular nitrogen ions which produce the nitrogen band have a lifetime of less than one-millionth of a second. In such a short time, the temperature of the molecule cannot change so very much from that of its parent gas. Therefore, these measurements should also give reliable temperature values for the atmosphere at auroral altitudes.

Before rockets and satellites were used in atmospheric and auroral research, it was this observational method which gave temperature information in the height intervals where the northern light occurs. The average temperature in the height regions from 100–150 km (i.e., auroral altitude) varies from about 200 to about 700 Kelvin.

7.5 Professor Kristian Olaf Bernhard Birkeland – A Pioneer in Auroral Research

a) Birkeland Introduced His Auroral Theory and Produced an Artificial Northern Light. Norway came into the picture a little late but nevertheless moved rapidly into the organized scientific research community which had bloomed in Europe in the 19th century. The man primarily responsible for this forward surge was Kristian Olaf Bernhard Birkeland (1867–1917).

After a few years of teaching and later studying in Bonn, Geneva and Leipzig, Birkeland was appointed a professor at the University of Oslo in 1898, when 31 years old.

Prior to this appointment, in 1896 Birkeland had introduced and proven his auroral theory. The main point of this theory was that electrically charged particles ejected from sunspots on the solar surface are captured by the Earth's magnetic field and directed

Fig. 7.11. Kristian Olaf Bernhard Birkeland (1867–1917) Norwegian professor in physics. He was originally devoted to mathematics but later became interested in physics. He was appointed professor at 31 years of age and introduced modern experimental physics to Norway

along the magnetic field lines into the polar regions. As the incoming particles reach atmospheric heights they are slowed down by the increasing density of atoms and molecules and in the process the atmospheric constituents become excited and ionized. In support of this theory, Birkeland performed extensive laboratory experiments to illustrate his points.

It was really amazing that Birkeland could actually demonstrate his new theory. The basic idea of these experiments was to study the motion of electrical particles (electrons) in a magnetic dipole field where the air density is low. The experiment – a model of the conditions in the Earth's upper atmosphere – in many ways reminds us today of how a television picture is produced by energetic electrons striking a phosphor screen. At the turn of the century, Birkeland's experiment was considered to be a very advanced idea in experimental physics. Due to the fact that the particles are electrically charged, they can be deflected by the Earth's magnetic field in such a way that they arrive on the nightside of the Earth. Birkeland's basic model experiment, in which for the first time a northern light was produced in the laboratory, was repeated by a host of scientists in other industrialized countries.

Birkeland's model experiment (Fig. 7.12) was in reality rather simple. In a large (about 1.5 m^3) vacuum chamber, he attached a little sphere, and in the sphere he placed an electromagnet. This vacuum represented space in which the Earth and the Earth's magnetic field were included. He then shot clouds of electrons toward this simulated Earth and thereby produced a light phenomenon which looked like the northern light. He could see that bunches of electrons were curved down toward and around the Earth's poles. The model experiment illustrated and proved his theory that the northern light was caused by electrons projected from the Sun. The electrons passed through space and were captured by the Earth's magnetic field into spiral tracks about the lines of force; they were guided into the Earth's nightside and in the process of slowing down due to collisions with the gases in the neutral atmosphere, northern light was created. Birkeland also thought

that with his so-called "terella experiment" he could demonstrate phenomena such as Saturn's rings and zodiacal light. In 1913, he prepared an extensive cosmological theory "about the world's origin", based on many of his laboratory experiments.

b) Birkeland's Expeditions. A cloud of free energetic electrons can interact with an electric current in the upper atmosphere because an electric current has its own magnetic field. Through association and logical reasoning, Birkeland now produced a new important theory, namely: the same particles which produce the northern light are also mainly responsible for variations in the Earth's magnetic field observed during such displays. In a popular way, one can say that he cleared up the question of why a compass needle is disturbed when northern light occurs. In order to prove his theory, Birkeland organized three expeditions (1897, 1899–1900 and 1902–1903) to the polar regions to gather the necessary observational data (Chap. 11.2). In addition he asked other observatories to be alert during his expedition periods.

From his observations, Birkeland computed the electrical current which flows in the upper atmosphere (ionosphere). He also brought forth a new

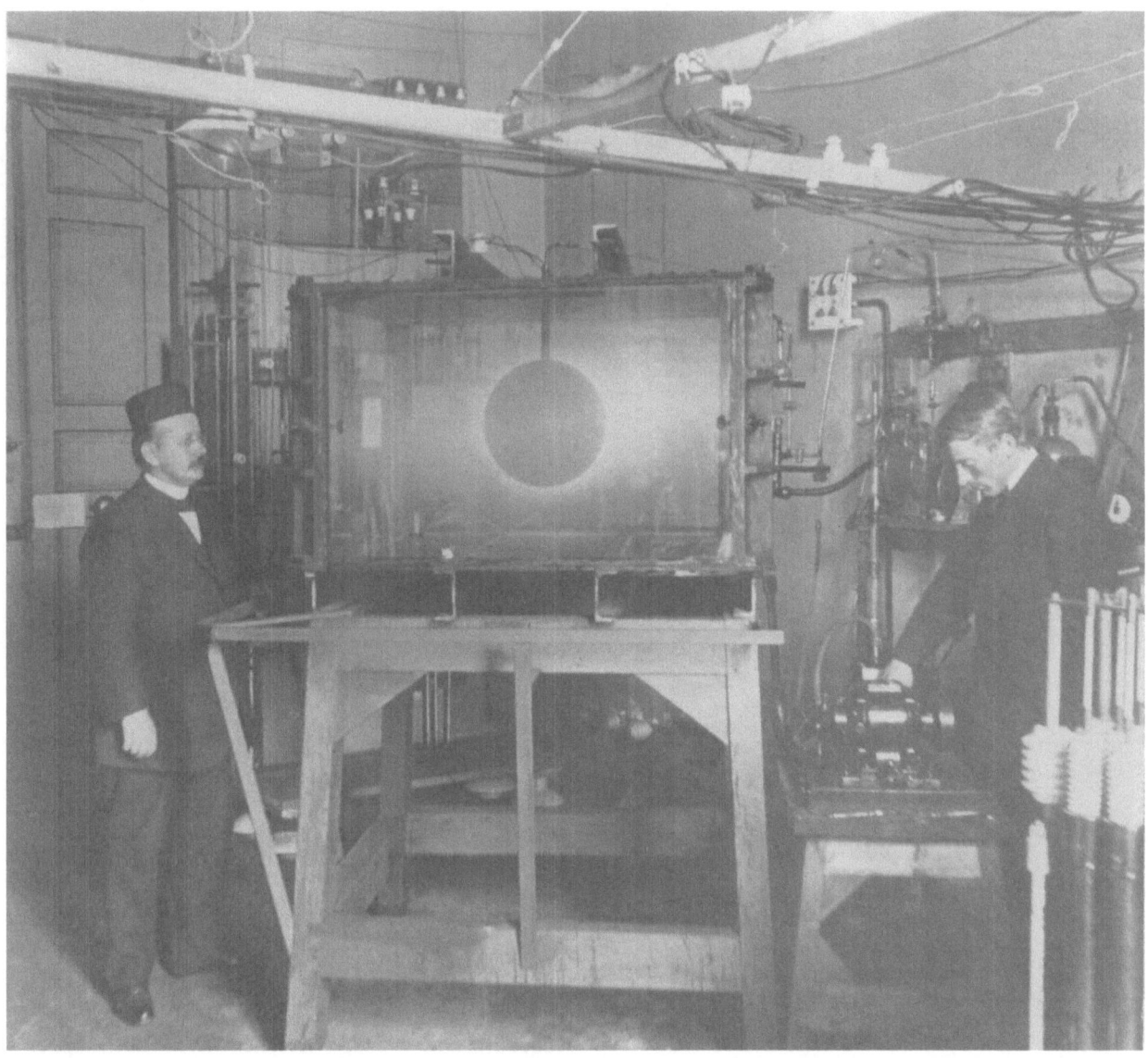

Fig. 7.12. Birkeland made his own "space" in the laboratory to create artificially northern light. He is shown here with one of his most devoted assistants, K. Devik. The Earth is represented by the "terella" which is surrounded by a glow

revolutionary theory for the magnetic storm which he thought is produced when an electric current flows along the Earth's magnetic field between space and the Earth's atmosphere. At auroral altitudes, the current curves away from the magnetic field and follows instead the auroral arc. The electric current along the Earth's magnetic field now generally goes by the name Birkeland currents.

He introduced the concept of elementary magnetic storms which nowadays are better known as auroral substorms, and he described the geophysical processes taking place in relation to the auroral dis-

plays. Birkeland also derived the so-called equivalent overhead current vectors which can be deduced from observations of magnetic field variations on the ground. The equivalent current system constructed in this way is, as Birkeland so correctly pointed out, only one of many possible solutions to this problem. The concept of the equivalent overhead current vectors has later been widely used in the study of the geomagnetic field variations caused by ionospheric currents. Not everyone has used this method with the same understanding as Birkeland did, and many premature conclusions have been drawn due to lack of insight into the limitations of the method.

Birkeland's theory of the currents associated with the northern light was too fantastic for the experts who lived during the first part of this century, and many of them were very reserved in their opinions.

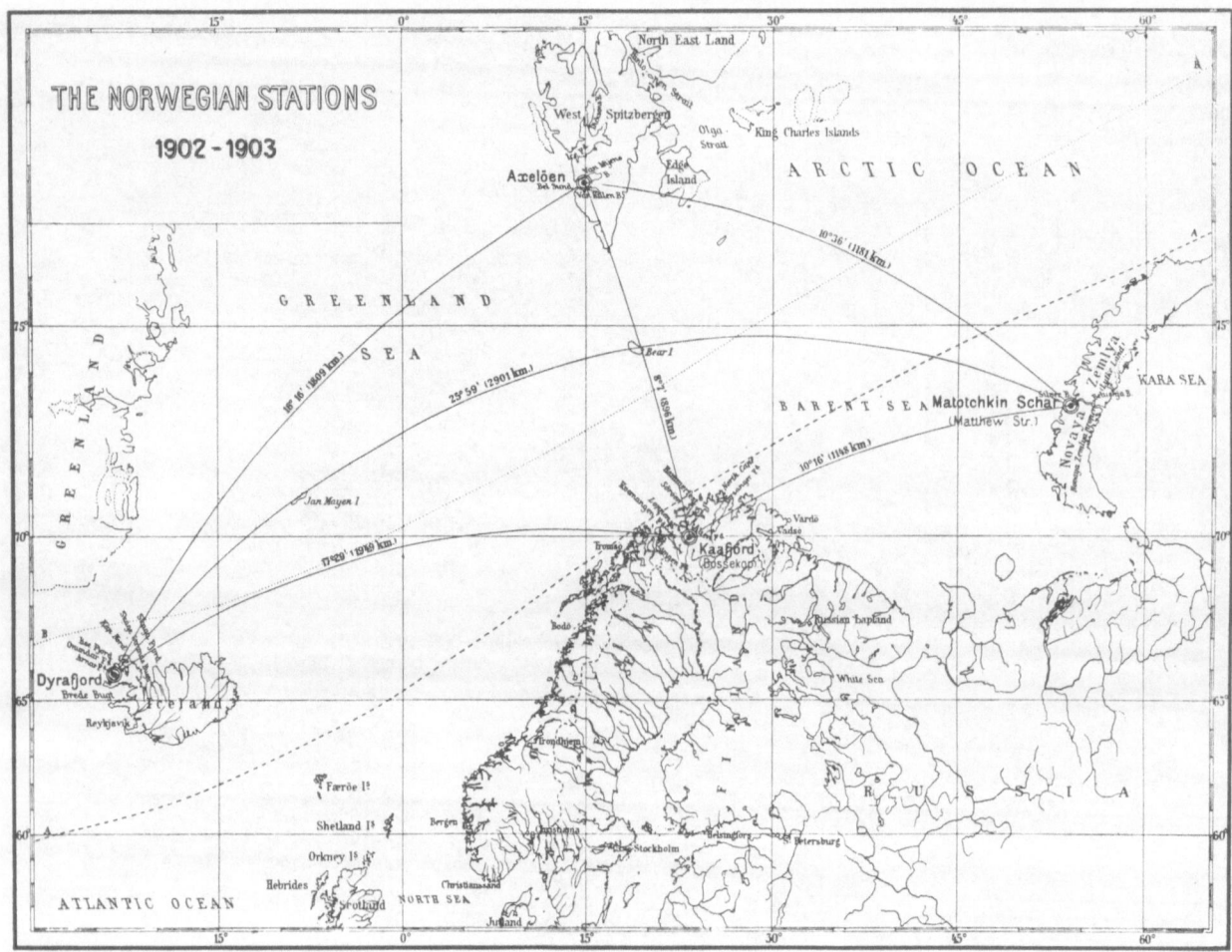

Fig. 7.13. A map showing the position of the four stations Birkeland equipped for his great polar expedition in 1902 in order to study magnetic variations in relation to northern lights

This reservation was probably due to the fact that Birkeland based his studies of the ionospheric currents on so-called event studies rather than on a statistical data set which was much more widely accepted. There is no doubt that both methods supplement each other, and there were thus no good reasons to disregard Birkeland's work.

Since the immediate space around the Earth has now been surveyed in detail with satellites and rockets, the attitude toward Birkeland's theory has turned to one of great admiration and almost total acceptance. Birkeland's work produced a new reality and interest and it has clearly shown "that one could again find gold in old gravel". His auroral theory is now accepted and in the last 20 years has been further expanded in a series of dissertations on the subject.

c) Birkeland Stimulated Scientific Research in Norway. Birkeland introduced modern physics in Norway and began a new research field in Norwegian science and cosmic geophysics, i.e., physics of the upper atmosphere. In a short but very hectic period he began a large number of new research projects and built up an active environment in the ancient Domus Media at the University of Oslo. He attracted to himself a small group of energetic young physicists who he inspired and stimulated to do their very best. It is worthwhile to mention here the names of some of these young physicists: Størmer and Vegard (from about 1904) and Krogness and Devik (from about 1906) (Chaps. 11.3 and 11.4).

Birkeland's work was extensive with a breadth and high caliber which had been lacking in earlier Norwegian research. He introduced innumerable ideas and theories; some totally general and revolutionary while others were utterly worthless. He had a creative and lively imagination and an enormous faculty for intelligently putting pieces of information together to obtain the correct solution to problems.

His principal contribution was nevertheless his ability to stimulate and inspire his co-workers which resulted in the golden age of Norwegian physical research which progressed in Norway thanks to Birkeland.

Birkeland's contribution as a university teacher was often surprising to the students. Olaf Devik appropriately describes this in his book *Amidst Fishermen, Physicists, and Other People*:

> "Birkeland had little time for lectures, but when he occasionally lectured on a subject in which he was interested he brought a breath of fresh air into the classroom, in contrast to the usual staid classroom atmosphere. Thus, he operated scarce electrical lecture equipment far beyond its rated capacity and burned out 100 Amp fuses with dignified nonchalance. Then he would stop and in a royal manner untie the ruffles in the sleeves of his jacket and dry his glasses in order to improve his view of his last miscalculation on the blackboard".

d) Birkeland Built an Electromagnetic Gun Which Gave him the Spark that Norsk Hydro Needed.

Birkeland also made considerable contributions to practical and applied physics. Most Norwegians associate Birkeland's name with the Birkeland-Eyde method for the production of potassium nitrate. Although this was an impressive accomplishment of considerable commercial value, it was merely one episode in the bountiful life work of Birkeland.

A fortuitous observation led Birkeland to build an electromagnetic gun, and in 1901 a company named Birkeland's Firearms was established. The main purpose in establishing the company was to build guns which could be sold to the military for money to finance auroral research.

A long straight electrical solenoid was used as the gun barrel and the projectile (with an iron core) was accelerated by means of a strong electrical current. The gun projectile weighed 10 kg. Birkeland considered this experiment analogous to "Münchhausen's cable" which Münchhausen cut off at the bottom and added to the top so as to climb to the moon.

Birkeland relates how the official demonstration ended when he used an extra large model of the gun:

> "It was at the University's old banquet hall on 6 January 1903. The gun was placed in the hall and pointed toward the target which was a three inch thick plank of solid wood. I had closed off the space on both sides of the projectile's track, but except for this area the hall was full. In the first section of seats were representatives from Armstrong and Krupp, the large weapon forging firm in Europe. I went through the principles on which the gun was based and "Ladies and Gentlemen", I said, "you may calmly be seated. When I push the switch on, you will neither see nor hear anything except the bang of the projectile against the target". With this I pushed the switch on. There was a flash, a deafeningly loud hissing noise, a bright arc of light caused by three thousand amperes being short circuited, and a flame shot out of the mouth of the gun. Some of the ladies shrieked and a moment later there was panic. It was the most dramatic moment in my life – with this one shot, I shot my stock down from a value of 300 to zero. But the projectile hit the bulls eye".

The comic paper *The Viking* was not gracious on this occasion but Birkeland took the misfortune with graceful humour and self-irony. This unsuccessful demonstration illustrates his most distinctive characteristic. His creative imagination immediately saw the meaning of the unexpected result of this experiment. This enormous, artificially created lightning served as the idea which created Norsk Hydro's nitrogen fertilizer industry. The first construction was begun at Notodden in 1907, and Birkeland's financial situation improved. He continued to spend a large part of his time in research. He had been hard of hearing since his experiments with radio waves in 1895 and was always in the forefront for "artificial devices" such as "trim and exercise" to keep himself in physical shape. It is therefore only "a half truth" when it is asserted that he travelled to Egypt to study the zodiacal light. He also made the trip for reasons of health. Birkeland died in 1917 before he was 50 years of age. It is ironic that at this time a working committee was in the process of nominating him for the Nobel Prize in physics.

e) Birkeland's Experiments Inspired Størmer to Calculate the Paths of Auroral Particles.

In 1927 Professor Størmer wrote:

> "It was Birkeland who used his ingenious insight into nature's secrets, which hide themselves in fascinating auroral phenomena, and he who created the experiments which led to the long row of works by himself and by his followers, in order to solve the riddle of the northern light. The story about the solution of the mystery of the northern light should certainly be of interest to Norwegians, because they can, without exaggeration, see that Norway here has led and still maintains the leading position in auroral research".

Størmer was appointed as a new professor in mathematics in 1902 and his great fascination for the natural sciences led him to be captivated by Birkeland's work.

After his introduction to Birkeland's terella experiment in 1903, Størmer immediately began a comprehensive mathematical calculation to determine how auroral particles can go from the Sun, through space, and into the Earth's atmosphere. Because of

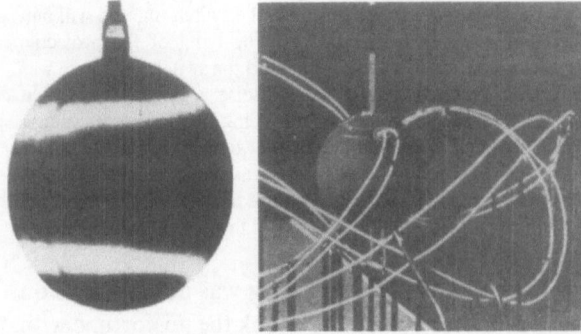

Fig. 7.14. To the *right* is shown Størmer's mathematical model of the trajectories of the solar particles entering the polar atmosphere, and to the *left* is Birkeland's "terella" experiment showing two parallel illuminated bands encircling each of the poles. Together these two models show a nice agreement between Størmer's and Birkeland's theories

the Earth's magnetic field and the electric charge which these particles possess, they spread out in an entirely different manner than solar rays of light, and the problem was therefore mathematically complicated. Here follows a short passage from the abstract to Størmer's first paper in 1904:

> "Birkeland has described an artificial aurora produced in the laboratory. From a theoretical viewpoint there exists an especially interesting problem – namely solving the equation of motion of an electron in a magnetic field. It is very clear what importance the solution to this problem will have to Birkeland's theory".

It was rather simple to set up the general mathematical equations which described the paths of auroral particles. The main problem was that these equations did not have a general solution (in the mathematical sense). Therefore, he was forced to use a numerical method and to follow each electron path step by step with tedious calculations.

Størmer's mathematical treatment of the auroral problem was a natural result of his education and talent as a mathematician. Altogether, Størmer and his assistants spent 18,000 h on these calculations.

Through a series of treatises he gave a survey of the possible solutions to this problem.

These papers, which were pioneer works in both physics and astrophysics, are now classics. It was shown how auroral particles from the Sun are sucked towards the Earth's magnetic poles in two circular zones – the auroral zones.

In this theory, Størmer also undertook calculations to determine why the northern light moves equatorward during increased solar activity. The principal idea was that particles from the Sun could also be guided by a varying electric ring current in the Earth's equatorial plane, located some Earth radii above the Earth's surface. This was also an important and correct step toward a solution of the auroral problem.

In the period 1920–1930, Størmer's analysis of an electron's motion in near space took on new significance and awakened an interest which went far beyond its original area of usage. This theoretical work was basic for the experimental demonstration of cosmic rays which were first discovered in 1925, i.e., about 20 years after Størmer began his work on this problem. A great interest was awakened in connection with mapping particle belts in near space, those which later became known as the Van Allen zones (Fig. 10.7) and which were observed by means of the first Earth satellites.

8 The Northern Lights as Weather Signs – and the Auroral Sound

8.1 The Northern Lights –
A Sign
in the Sky of Changing Weather

Since time immemorial the human being has been strongly influenced by mysticism. We could well ask ourselves, even today, who among us does not harbour at least one small superstition in the depths of his subconsious being? Within the general category of superstitions lies a group of feelings, thoughts, or ideas which are labelled "signs" – a sign in this context taken to mean an indication of things to come. One such "weather sign" which is familiar to English-speaking seamen is "red at night – sailors delight, red in the morning – sailors take warning". Here, the reference is to the appearance of the Sun when it rises and sets.

For people who obtain their living from nature, whether from products of the soil or creatures from the sea, weather is a decisive element. In earlier days any sign, in the sky or on the Earth, among fish or birds, creatures or vegetation, which could be related to the weather for tomorrow, was closely followed and studied. In the following we explore how the northern light influenced people in their thoughts as to how the weather might be in the immediate future as well as for a whole season.

The northern light is a phenomenon which people often believed gave a portent of weather and wind. In Scandinavia today there are still untold "weather signs" connected with this sight in the sky. Indeed it has been such common practice to predict the weather from auroral displays, that even today the northern light is called "weatherlight" in many places in Norway.

Local weather signs, both in the old days and today, are strongly influenced by individualism and the ingenuity of the soothsayer. As an example of the importance some people put in such weather signs we quote a few words from a story told from Lesja in Norway by the well-known astronomer Sigurd Einbu (1866–1946).

"The northern light was an omen of storm and bad weather. People believe that still today. Not long ago a man gave me an example of this. One morning he warned the boys not to go to the mountain to gather moss since great northern lights had been seen the night before. I tell you they learned to respect the northern lights! It was such bad weather during that day that they only just came back home that evening and with an empty sledge".

Each district or neighbourhood has its own signs and in a few cases one can find good agreement about the weather signs over large areas of the population. In Hattfjelldahl in Northern Norway people believe, for example, that if the northern light moves far to the south in the sky, the weather will turn milder, and this will especially be true if the northern light is bright. At Beitstad, about 225 km north of Hattfjelldal, people expect precipitation if the northern light has a similar position. If, in addition, the northern light has a red colour, the precipitation will be extra strong. Considering the fact that these two places are only 225 km apart, and that the same display can be seen in Hattfjelldal and Beitstad, the weather signs are not directly in opposition to each other. During the winter months, precipitation often follows when the weather becomes milder, and after cold or clear nights milder weather eventually must follow.

Some of the most detailed weather signs related to the northern lights are probably those found in *Description of Søndfjord* by Hans Arentz (1731–1793), from which we quote the following:

"But the most common and, maybe due to the many circumstances in which it shows up, that possessing the strongest prophetic powers and among the most beautiful appearances in our sky is certainly the Northern Light, which stays away only on a few winter nights, and indeed it can be seen in the clearest moonlight ...".

And Arentz later continues:

"When it (the northern light) comes to a standstill with a brilliant clearness, or makes an archway between northwest and southeast without any considerable motion, one maintains, and that is not completely uncertain that the weather conditions will within some time be clear and cold, if on the contrary the northern light spreads out across the sky toward the south and west with radiating arcs, people forecast windy weather, and the more rapidly the rays are playing, igniting and extinguishing, soon at another place of the sky, the stronger wind is expected".

The weather signs relating to the northern light in Kåfjord, Finnmark, were also very detailed. People thought that if the northern light was observed low on the horizon to the east, the weather would be dry and cold; on the other hand if it was seen low on the horizon to the west; then storm, snow and mild weather was to be expected. In the Gudbrandsdalen valley the northern light was a sign of clear weather, but if it was whitish it indicated snow. In Sunnmøre, on the west coast of Norway, people believed that the northern light was an omen of changeable weather from clear to squalls and precipitation.

8.2 The Northern Light was Often Called "Windlight"

Many years ago in most places in north Norway people expected windy weather when the aurora appeared and "windlight" was also a fairly common name for the northern light among the Norwegians.

It was also common in Finnmark – even at the beginning of our century – for fishermen to call the northern light by the name "vindlys" or "windlight". If they had been watching a very strong display in the evening, they rowed out fishing the next morning with anxiety because they then expected a strong wind without any preferred direction. There was a strongly related belief among the people in Hardanger on the west coast of Norway, where they thought that the fish would not bite well during auroral events.

In the weather signs from Hedmark in the southeastern part of Norway and in the western part of Norway it was also common to relate the northern lights with windy weather. In large areas of the country, however, wind from the south was most commonly expected if the northern light moved south of the zenith; the belief being that the wind tried to blow the light back to its primary home – namely in the north.

If the northern light was reddish it was often called "brag" or "bragd" and in this respect words like weather-"brag" and wind-"brag" were used. The meaning of "brag" is something that flares, flickers, or glints. In Valdres in central south Norway, people thought that it was stormy in the North Sea when the "brag" was seen; as if the rolling icebergs were reflecting light into the sky. The idea that there is a connection between red northern light and windy weather is probably a very old one. This is readily seen in a book by the sea captain Johan Heitman from 1741 (see Chap. 5.4)[17].

"Yet have seamen commonly experienced in the north as well as also fishermen living along the coast of Norway, that sign in the northern lights when it is seen in the west that they expect a southwesterly wind, and thereby is the opinion of the versed in natural history verified, that the air in a similar way has its several saline constituents, which stir up the air at certain times, as the ocean has its own. In spite of the fact that such prognosticated variabilities in the weather most often occur, it happens in some cases that some time elapses before the change arrives; yet it is certain that the cold regions in the air contribute much to the variability in the weather and the force behind it; by preference when the northern light is seen like red copper, then comes certainly enough an impetuous storm from the west – and northwest, yet the weather can shape itself sailable for a week or more thereafter, then the storm comes first, of which I have had many examples".

Heitman – practical sailor as he was – hinged his arguments on his own experiences.

As a curiosity, it is interesting to note that Heitman argued that the northern light would be less frequent in the colder winters, because the air would then be heavier and subside. The fact that the northern light was less frequently observed in the 17th century, Heitman maintained was related to the cold weather (the Little Ice Age) which was then dominating.

8.3 The Northern Light as a Weathersign in the Northern Countries

It is worth noting that in many of the numerous topographical descriptions of the Nordic countries made in the 18th and 19th centuries, the northern light is almost always mentioned together with the climate. This is not accidental since in the north it was commonly believed that the northern light was a phenomenon occurring in the lower air and sometimes even below the clouds. It is therefore not surprising to find, that throughout the Nordic countries, weather prediction based on auroral displays has been widespread.

In Hålogaland in northern Norway it was common to make long periodic weather forecasts based on the activity of the northern light. Many people in this region thought that if the northern light was ex-

---

[17] As this is written in Norwegian from the 18th century we have tried to keep to the characteristic style in the translation

aaned.	Thermometret			Barometret			Klart. Dage.	Taaet eller Ponti.	Regn eller Snee.	Stille.	Vind.	Storbyt.	
	Medium	høiest	lavest	Medium	høiest	lavest							
Septbr.	+6,03	+10,3	+0,8	27"10,55	28"5,4	27"0,2	13	11	6	12	18	3	høieste Temp. 9de
Octbr.	÷0,16	+9,8	÷10,4	27"9,80	28"3,3	27"4,0	13	13	5	13½	17½	6	og laveste den 15de
Novbr.	÷2,50	+5,7	÷14,1	27"9,41	28"4,5	27"3,1	15	11½	3½	4½	25½	5	Decb., forste Snee
Decbr.	÷8,40	+2,2	÷17,3	27"8,40	28"4,6	27"0,9	17	12½	1½	5	26	6	11te Decbr.
Januar	÷4,31	+5,5	÷16,9	27"9,80	28"3,5	26"1,0	10	15½	5½	5	26	10	Begyndelsen af
Febr.	÷0,01	+5,1	÷13,1	27"7,93	28"4,9	26"11,1	5½	15½	8	4	25	5	Jan. Snee næsten borte.
Marts	÷2,00	+5,0	÷9,3	27"5,32	28"1,3	26"9,1	6	16	9	3	28	4	
April	+2,15	+7,6	÷5,7	27"7,22	28"4,3	26"9,4	7	15	8	9½	20½	—	
Mai	+3,63	+11,3	÷5,1	27"11,00	28"4,5	27"3,8	11½	16	3½	7½	23½	—	laveste Temp. 1ste
Juni	+7,62	+15,9	+2,5	27"11,60	28"2,8	27"5,7	12½	13½	4	4½	25½	—	Mai, høieste den
Juli	+9,14	+17,4	+4,0	27"9,06	28"1,0	27"4,2	11½	13½	6	4½	26½	—	30te Mai.
August	+6,79	+14,1	+0,8	27"11,24	28"3,3	27"8,0	15	10	6	6	25	—	
ele Aaret	+1,°52	+17,4	÷17,3	27" 9,28	28"5,4	26"1,0	137	163	66	79	287	39	

Fig. 8.1. Government officials appointed by the king in the 19th century commonly took meteorological notes. One such civil servant was the pastor Fredrik Rode who served in the district of Talvik in Finnmark from 1826 to 1834. During one year from September 1831 to August 1832 he made his meteorological notes and organized them in the present table. In the last column he also added the number of days with northern lights per month. Rode did not reject the idea that the northern light could be a token for wind and stormy weather; characteristically enough he has placed the number of windy days next to the column of days with northern light

tremely lively or occurred frequently during the autumn, then the coming winter would be cold.

In Sweden a similar prediction was found which said that

if the northern light shows up early in the autumn the coming winter will be severe.

Another well-known prediction was that six weeks after the first northern light in the autumn, winter would be coming.

In Finland the northern light was an omen for weather changes, and for people in Karelen it was a warning of cold weather and snowstorms.

In the Faroe Islands the following weather signs are known:

When the northern light is standing low
it will be fair weather.
When the northern light is standing high
it will be bad weather.
When the northern light is uneasy
it will be windy.

In Jokkmokk in Northern Sweden, one finds the following tie-in with the northern lights, which is similar to the incantation found among the Norwegian Lapps (Chap. 1.2):

Northern light, lip, lip, lip!
The old greybeard's butter trowel
is catching up with you. Lip, lip, lip!

Here the butter trowel is probably the crescent moon. Since the moon is an omen for cold weather and the fluttering northern light presages mild weather, it was important to the Lapps in Jokkmokk that the northern light be in such motion that the Moon cannot come and dazzle it with her light.

A half poetic weather sign from Hvaler in the southern part of Norway is more in the form of a statement than a prognosis, and reads as follows:

A swarm of northern light
in the heaven high
is a token of a clear sky.

The northern lights were also used as weather signs outside Scandinavia. The Creek Indians in Georgia, Alabama, and the Cheyenne Indians of Wyoming and Colorado, said that the appearance of the northern light meant the weather would change for the worse.

The Penobscot Indians of Maine credited the northern light with bringing a windy day. If the lights were flickering during the display, the wind would blow strong and steady; if they were still and quiet, the wind would be squally.

In Massachusetts, even into the 19th century, the saying was that:

South wind and storm will come within 24 h after northern lights.

Scientists today have not found any conclusive evidence for a relationship between the northern lights and the weather, except that obvious clear weather is needed for the human eye to see the light from the ground. It is surprising to notice the similarity in weather signs regarding the northern lights across the northern hemisphere. Of course some of the signs are often a matter of course, like the prediction saying that milder weather will come after an auroral display. If the northern light can be seen, the weather must be clear and then it is usually cold; after clear weather it must be cloudy and then the weather turns milder. The detailed forecasts of wind directions in association with the position of the northern light is of course far more speculative, but even so it is astonishing to see the similarities in detail in these forecasts between the Indians in North America and the Scandinavians in North Europe.

8.4 The Northern Lights Used as
a Weather Prediction –
A Subject of Current Interest for Discussion

Sophus Tromholt (Chap. 6.5) was one of the first to make a thorough research among the different weathersigns connected to northern lights. In his book *Under the Rays of the Aurora Borealis* he writes the following about this matter:

"I will by no means deny that there is a Relationship between the Northern Light and the Weather, on the contrary, I consider it even probable; but this connection is not as straightforward, that it can simply be proven by the Help of a few Years supposed Experience, and furthermore this Problem has so far not become the Object of a real scientific Exploration. It must however, be considered likely, that some Weather Conditions can be more propitious than others for the Formation of the Northern Light, and maybe also, that some Kinds of the Northern Lights can exercise some Influence or at least have Bearing upon the coming Weather".

Tromholt wrote to a large number of government officials in the Nordic countries and asked them to give their opinion on the northern light as a weather prognostic. He presented the result of his inquiries in the following manner:

"As could be expected, the Opinions are very divided; nevertheless it cannot be denied, that there is a fairly good Agreement about the Couple of Points, which therefore cannot be rejected without further ado, but on the Contrary, Significance must be attached as a Result of demotic Tradition and Experience. Most of the Enunciations are in complete Accord with Regard to the Point that low standing Northern Lights in the North are related to steady, cold Weather and Snow, while the Northern Light in the South heralds mild Weather, southerly Winds and Rain, and furthermore that a Break in the Weather and strong Blowing are accompanying intensely flaming Northern Light".

Much has been written in the literature, but of variable quality, about the relationship between the northern light and different cloud formations. Both the forms and colours of different types of clouds can be very similar to the form and colours of certain types of northern light. In the Nordic countries it was Christoffer Hansteen (Chap. 6.1) who, during the last century, related the northern light to different types of clouds and weather conditions. His idea was based on a firm belief that an auroral substance emerged from the ground to the upper atmosphere, and acted as condensation sources for the moisture in the air as well as causing the northern light. Based on such a theory it was natural to expect a relationship between the formation of clouds and northern lights. Even Kristian Birkeland (Chap. 7.5) when planning his grandiose expedition to the Haldde mountain in Finnmark in 1910, applied to the government for money with the following arguments relating the northern light to weather conditions and clouds:

"During my last expedition up to the present Haldde observatory I found an electrical linkage between the earth's magnetism and meteorological phenomena; as the air shows up as strongly ionized after gigantic magnetic storms.

Such an ionization has a very strong influence on the formation of the clouds and on the electrical conditions of the Earth. In full agreement with this stand my previous experiences which seem to show that observations of magnetic variations at high latitudes will be of importance to the preparation of weather forecasts.

I therefore believe, that a permanent observatory at Haldde will be an El Dorado for scientific discoveries which can be of advantage to meteorology".

Also in his famous work *On the Cause of Magnetic Storms and the Origin of Terrestrial Magnetism* (cf. Birkeland, 1908) he devotes a chapter to this relationship and writes:

"There was another phenomenon, striking examples of which we had opportunity of seeing on this expedition in May 1910, namely the formation of what may be called auroral clouds. In addition to the usual polar bands, which in a clear sky, could very often be observed in the form of several evenly luminous arcs, of which, however, *one* was especially conspicuous, exactly similar to parallel auroral arcs, we very frequently found formations of cirrus clouds, which exhibited the most perfect agreement with various auroral formations".

Fig. 8.2. The Norwegian professor Carl Størmer tried to explain the relationship between auroral light intensity variations and variations in auroral sound as reported by many observers, as being due to electrical discharges on the ground. Here Størmer is seen working with his camera at Talvik in Finnmark together with his assistant Bernt Johannes Birkeland. This picture has been widely distributed with an erroneously given name for Størmer's assistant

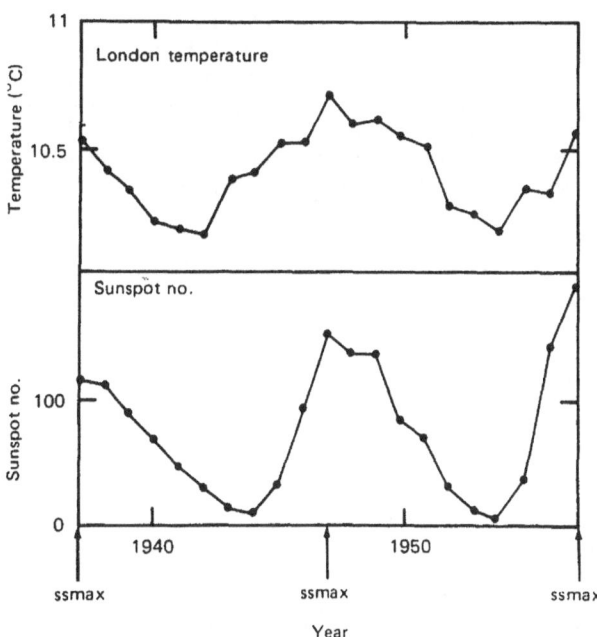

Fig. 8.3. The relationship between the solar activity and the weather has often been discussed in the past. A connection seems to exist statistically, but no satisfactory physical explanation has been found. The above shows, however, that the average air temperature in London between 1935 and 1957, i.e. for almost two complete solar cycles, follows the variation in the sunspot number for the same period (King et al., Nature, 1974, 252, 2)

Birkeland also noted that probably the first observer to point out this relationship was Adam Paulsen (Chap. 6.7). By far most auroral displays arrive on the scene in the thin air located about 100 km above the ground. Since the weather is mainly determined by air conditions below 10–20 km above the Earth's surface, a direct physical linkage between the northern light and the weather is very unlikely.

8.5 Satellite Observations Have Again Brought the Relationship Between Weather and Northern Light into Focus

Recent satellite observations have, however, raised discussions on the northern light being a "weather light". Some very recent data indicate that if a very strong auroral disturbance takes place above the North American continent and lasts for several days, low pressure will develop more easily in the western Atlantic and in the Greenland area, than during periods of no such auroral disturbance; and furthermore *these* low pressure areas also have a tendency to strengthen, relatively more than others. It is well known that the low pressure areas developing close to Greenland have a very strong influence on meteorological conditions in Scandinavia. Therefore rather than taking the auroral displays above our heads as a token of the coming weather, maybe one should follow the development of the northern lights seen above the North American continent.

The first, preliminary results of these new explorations, have lead to an increased interest in the relationship between the northern light and the weather during the last 4–5 years. If there is some physical interaction between weather and the northern light it must be very complicated indeed, but more and more auroral researchers are beginning to believe that there is a reciprocal influence. Meteorologists, however, are essentially doubtful of an existing influence. Most likely, if there is an interaction, it is meteorological conditions that can be important to the occurrence of the northern light and not vice versa. This

is because the total amount of energy involved in the meteorological system is of the order of million times larger than the energy connected with an auroral disturbance. If the auroral disturbance should then have an influence on the weather it would have to be through a kind of a catalytic process. Within the coming years we will probably learn more about these interactions, but what the future will bring in this respect is probably as difficult to predict as the weather itself.

8.6 Can the Aurora be Heard?

A very controversial question concerning the northern light is whether or not it can be heard by human beings. This is a question that has been discussed for several centuries, and the discussions still echo in some hidden corners of the scientific world. One positive result can often have a stronger impact than hundreds of negative results, and many well known auroral scientists have officially taken very strong points of view on this subject.

It has been suggested that the first reference to auroral sound was made by the Roman historian Tacitus in his book *Germania* from 98 A.D. It is, however, rather doubtful that the sound mentioned by him has anything to do with auroral sound, but is more likely related to the sound our ancestors believed to be caused by the break of day.

The Swedish physicist Torbern Bergman (Chap. 5.15), however, in the 18th century measured

Fig.8.4. The American professor Elias Loomis (1811–1889) is probably the most forgotten and overlooked auroral physicist in the 19th century (Chap. VI.6). In addition to composing on a statistical basis the very first auroral zone he also discussed the reality of auroral sound. His explanation based on psychological arguments is remarkably similar to the modern view of this audio nervous response to a violent light phenomenon

the height of the northern light as somewhere between 380 and 1,300 km and from this he argued very convincingly against any audible auroral sound being heard on the ground. He claimed that since the northern light takes place at such a high altitude where the air is strongly rarefied, any sound created by this light cannot penetrate to the ground.

Christopher Hansteen (Chap. 6.1) in the beginning of the 19th century also discussed auroral sound, and was fairly convinced that this sound was audible at the ground. From Hansteen we quote:

"How can this, at such a height, produce a certain sound? From the manifold experiences we have had of this in the north, it must be considered as a fact admitting of no question".

The American physicist Elias Loomis (1811–1889) (Chap. 6.6) also discussed the likelihood of audible auroral sound and he had a very different opinion from Hansteen. He claimed the following concerning auroral sound:

"When we see a brilliant light shooting like a rocket across the sky, it is natural to expect an accompanying sound. People generally hear what they expect to hear. Tacitus informs us that the ancient Germans heard a noise whenever the setting Sun descended into the western ocean.

No observer has ever spoken of the interval that had elapsed between the darting of the auroral rays and the alleged noise. But, on account of the elevation of the aurora, this interval should be a long one. Sound requires 4 min to travel a distance of 50 miles. It is probable, therefore, that the sounds which have been heard during exhibitions of the aurora are to be ascribed to other causes than the aurora".

This argumentation, made more than 100 years ago (in 1866), can withstand any criticism of today's physicists. The psychological explanation that people hear what they expect to hear, is very similar to the modern view that widespread brilliant lights in rapid motion are expected to be accompanied by sound, and in fact that the violent impression of such light phenomena can indeed trigger the audio nervous system, and thereby give a false response of sound.

A large amount of data collected by active observers shows that many people have heard sound connected with strong auroral displays. The most common description of the auroral sound is rustling, hissing, whizzing and crackling noises as if somebody is wrinkling aluminium foil, the rippling of a distant river or waterfall or the sound from a sail or a flag flapping in the wind. Some people describe the sound to be like crackling in the snow, the voice of a blackcock, breathing and groaning etc. Furthermore, some people claim that the sound varies in accord with the variations in intensity of the visual northern light itself. Towards the end of the last century Sophus Tromholt (see Chap. 6.5) collected a great many reports on auroral sound. In the following we reproduce three of these reports:

"That I on the 20th and 22nd of January together with the 29th of March during my Observation heard Sound accompanying the Northern Light, is something, that is certain to my subjective Conviction. On the 20th January the Northern Light showed up so strongly, as one ever has the Possibility to see it in this district. It emerged in the northern Horizon, which almost always has been the case this Winter, like an Arc, but developed little by little into flaring flames, once in a while these spread out with an extreme Rapidity in beautiful white, yellow, red and purple Colours from the Zenith; it appeared that Evening, which I personally have more often observed, that the Northern Light remained at a closer Distance to the Ground than is common. The Sound was heard, as the Northern Light almost had reached Zenith; sometimes weaker, sometimes stronger as the Light moved itself. The Sound can to the closest be called a Crackling. The Weather in the Evening was quiet, the Sea calm, the Sky clear and the Air relatively mild. I stood for a long Time quiet and observed the Phenomenon from a little Elevation close to the Farm Tunstad. I do not think that any Illusion took Place. The 22nd of January, a similar beautiful Auroral Evening, I was enroute from Søsand through Gaasø and towards the Prostfjord, accompanied by Julius Herness, a Son of the rural

Major in this Place; arrived just outside Øksnos we stopped Rowing to, in Quiet, observe the Northern Light, which as on the 20th stood in Zenith and a little to the west of it. We both heard the Sound, and agreed that it belonged to the Nothern Light, it was of the same Nature as that on the 20th and appeared to decrease and increase as the Lightflames moved. On the 29th of March I observed the Northern Light at Tunstad under the same favourable Conditions, calm Weather and clear Air. The Sound was then also heard in a similar way".

"By the many Northern Lights, I in a long Series of Years here in Denmark have seen and almost always very carefully observed, I have been so lucky that I several Times have noticed such a Sound, the first Time was the 16th September 1838. It was, considering the Season, a fairly nice Afternoon, one of the few lovely Afternoons that the cold and wet Summer of 1838 offered us, when I was outside Copenhagen's Nørreport. It was between 10 and 11 o'Clock, cloudless Sky, and it was completely quiet. A beautiful Northern Light had the whole Afternoon been outspread across the northern Part of the Firmament. At that said Hour "suddenly" as I at that Time took down, "the whole northern Sky appeared to be in a wavy Motion, as manifold transparent, quivering Light-clouds with great Speed and with a peculiar quite never before heard by me, sizzling Sound went up towards the Zenith".

"Only once, in November 1856 at Beskades, a Mountain between Alten and Kautokeino, about 1,500 Feet above Sealevel, I believe during an extraordinarily gorgeous Northern Light with a brilliant Corona, I have heard a peculiar, feeble crackling Sound in the Sky. My Fellow-Traveller heard the same, and I can remember, that we paused and talked about this Sound".

8.7 Can Sound Propagate From the Auroral Altitude Down to the Ground?

The human ear is sensitive to sound waves with frequencies from 20 to 20,000 Hertz (Hz) (one Hz is one cycle per second). Sound waves are pressure waves which propagate at a speed of about 340 m per second in air at normal pressure (one atm) and normal temperature (about 20 °C).

In Fig. 8.5 the speed of sound is presented at different altitudes between the ground and 110 km. The average speed of sound in the altitude interval between 110 km and the ground is close to 300 m s^{-1}. This means that an audible sound wave would need about 400 s to propagate from this altitude to the ground. As light is propagated at a speed of 300,000 km s^{-1} it requires only about 0.4 ms to trav-

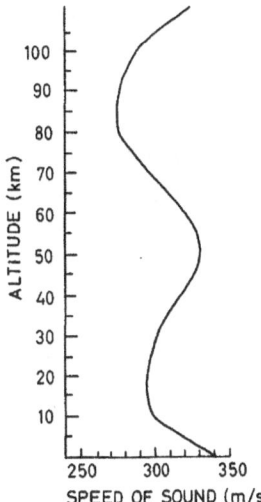

Fig. 8.5. The speed of sound varies with altitude as indicated in this graph. The average speed of sound between the auroral height and the ground is about 300 ms^{-1}. This means that a sound wave would need about 400 s to propagate from an auroral display down to an observer on the ground. The light emissions, however, would need only 0.4 ms to cover the same distance, therefore these two signals cannot vary in phase as has often been reported by numerous observers

el from an altitude of 110 km to the ground. This means that any variation in sound, if it is caused by variations in the northern light cannot be observed in phase with light variations on the ground but rather will be delayed by more than 6 min.

As mentioned in Chap. 10.2, air density at the heights of the northern lights is very low – only about one-millionth of the air density at ground level. At such a low density, the average distance between the collisions a gas molecule experiences in air (the mean free path) will be much longer than the wavelength of the sound wave, and therefore the sound wave cannot easily propagate but instead is strongly absorbed or damped.

This is illustrated in Fig. 8.6 which shows the altitude at which a sound wave must be dispatched in order that 10% of the wave energy can reach the ground. Whereas soundwave of 1 Hz can reach the ground with 10% of its energy if it is emitted from about 100 km altitude, an audible sound wave will not reach the ground with as much as 10%

Fig. 8.6. This graph shows at what altitude a sound source has to be situated in order that the sound waves from such a source can reach the ground with 10% of their energy. Sound waves originating above this curve will reach the ground with less energy. Audible sound waves with frequencies above 20 Hz will therefore be heavily damped above 50 km, and are not able to propagate from auroral heights to the ground unless the source is unreasonably strong

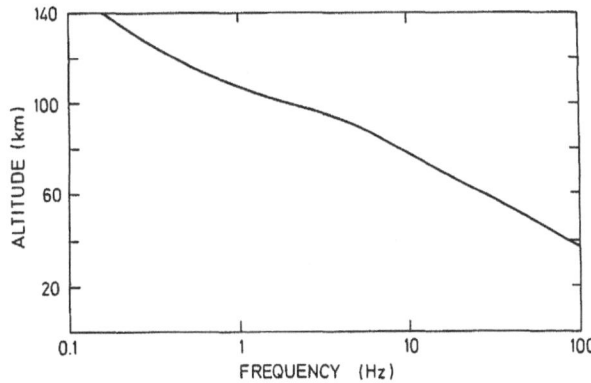

of its energy if it is emitted from more than 40 km altitude. For higher frequencies this height limit will be even lower.

At ground level an audible signal which can be heard over distances of 1 km will at an altitude of 20 and 100 km travel only 50 and 1 m, respectively. It is therefore inconceivable that audible sound generated at auroral heights can be strong enough to be heard on the ground after propagating through the atmosphere.

As already mentioned, many observers assert that auroral sound varies in phase with intensity variations in the northern light. This is possible only if sound is created at a distance less than a few meters from the observer. If the sound originates from larger distances of say a kilometer or more, there will be a time delay of a few seconds between the light variations and the sound variations. A simple explanation of the two phenomena observed in phase was given by Professor Carl Størmer (Chap. 7.1) in 1927. His hypothesis can be recapitulated as follows:

"A strong northern light generates a local discharge on the ground, and it is this discharge that can be heard as the crackling auroral sound".

For such a discharge to occur in the air close to the ground, an electric field of close to 1,500 V m^{-1} is required. That electric fields occur in the neutral atmosphere has been known for a long time, but these fields are only of the order of 100 V m^{-1}. At about 1970 it was found that the electric field close to the ground can reach several thousand volts per meter during auroral displays. From such an extraordinarily high electric field, discharges can well occur close to the ground, and such discharges may make a buzzing or a crackling sound which is often associated with the auroral sound. If such discharges occur close to the observer, no noticeable time delay would be expected.

Since many reports of auroral sound stem from experienced and reliable observers, they cannot be brushed aside as fallacies. No instrumental, objective measurements of audible sound created by auroral displays have so far been presented, in spite of several attempts with very sensitive microphones. In order to eventually test Størmer's hypothesis it will be necessary to carry out simultaneous instrumental observations of sound, light and the electric field at ground level.

9 Northern Lights and Geomagnetic Disturbances – Their Influence on Daily Life

9.1 The Northern Light – A Useful Light Source

The beauty and grandeur of the northern light is aesthetically satisfying to many people in much the same manner as any other beautiful work of art. To others it may be an awesome and fearful sight predicting disasters and dire consequences. It also has been incorporated into the folklore and mythology of many people living in the areas of strong auroral activity. As we have seen, many groups of people believe it to be a weather predictor.

There are other and more mundane influences of the northern light on the daily lives of people living in the auroral zone. Light from the northern light can be quite beneficial under certain working conditions – but the accompanying geomagnetic disturbances can be both bothersome and destructive. These more "down to Earth" effects of the northern light and geomagnetic disturbances are the subject matter of this chapter (see Chap. 8).

In his book *Optegnelser fra Finnmarken* (*Notes from Finnmark*) written in 1842 Fredrik Rode (cf. Fig. 8.2) belittles the idea that illumination from the northern light is an important light source in everyday affairs. We quote him as follows:

"– however one can yet from this conclude, that the Northern Lights in Finnmarken by no Means is so frequent, that they, like the Textbooks in Geography tell the Youth, could possibly make Compensation for the Want of Daylight. The Northern Lights are too indefinite and inconstant that it possibly could occur to somebody to reckon on them, or in the remotest Way consider them important when taking Decisions concerning ones Business...".

Many writers have made exaggerated claims (resulting in attacks such as that of Rode above) concerning the ability of people to perform tasks under auroral light. One example of such an overstatement is from a book by T. Haukenæs called *Midnattsolens Rige* or *The Land of the Midnight Sun*:

"The northern light is not only a magnificent and exalting sight in these northern areas, but it is also of great advantage for the inhabitants, as one can by the use of the auroral flames see how to carry out different kinds of work. Many times illumina-

tion by the northern light is as strong as faint moonlight night, and many times fishermen row to sea without any other light than that which they obtain from the flaming aurora borealis, like the Lapps who by aid of the same illumination wander around in the mountains looking for their reindeer; while those at home use the light to perform household duties."

Even though Haukenæs and many other authors exaggerate the usefulness of the northern lights, there are more accurate descriptions of the use of this light source. The first account concerning the northern light in this respect is in *The King's Mirror* (Chap. 2.2) and we quote:

"While these rays are at their highest and brightest, they give forth so much light that people out of doors can easily find their way about and can even go hunting, if need be. Where people sit in houses that have windows, it is so light inside that one can see each other's faces".

According to *The King's Mirror* the northern light could be of limited help to travellers as well as to hunters.

The difference in wording by these authors could have been influenced to some extent by the state of the northern light at the time each of them lived. For instance, Rode lived in Finnmark from 1826 to 1834 during a period of rather limited auroral displays, while Haukenæs wrote his book at the end of the last century at a time when the auroral activity was very high. The author of *The King's Mirror* who probably did not personally see northern lights, reports on what he had heard from travellers to Greenland. These travellers probably visited Greenland at a time when the auroral displays there were relatively numerous and strong.

It has also been claimed that Eskimos used the northern light in two different ways as a navigational aid, when they lost their way on the ice far away from home. If there was a strong auroral display it would illuminate the contours of the mountains and shorelines sufficiently to allow them to set a course for home. In the second case, since auroral arcs mainly are aligned in an east–west direction, it follows that magnetic north is the direction towards the highest point of the auroral arc. This fact could provide information similar to that of a rough compass as an aid in returning home.

Although we might think it rather unlikely that illumination from the northern light would be useful in hunting, we should point out that one of the first realistic drawings of such a light in Norway illustrates a hunting scene during an auroral display (Fig. 1.1) and drawings similar to this one are known from other countries (Fig. 9.1).

9.2 The Northern Light and Electrical Disturbances

Associated with auroral disturbances at auroral latitudes in the polar regions are intense electrical currents of up to 1 million Amp in the upper atmosphere (at heights about 120 km above sea level). The presence of these currents can be recorded on the ground with simple magnetic instruments [magnetometer (Chap. 11.7), compass etc.]. High electrical potentials are induced in very long metal wires by these strong currents. Metal layers in the Earth itself are similarly influenced by strong electrical currents at auroral altitudes.

In the early days of the telegraph and telephone it was well known that serious disturbances to these systems could occur when the northern lights were very active. One of the first to discuss this fact in a scientific paper was the American professor Elias Loomis (see Chaps. 6.6 and 8.6). In his article *The Aurora Borealis or the Polar Light* in 1866 he writes:

"Auroras exert a remarkable influence upon the wires of the electric telegraph. During the prevalence of brilliant auroras the telegraph lines generally become unmanageable. The aurora develops electric currents upon the wires, and hence results in a motion of the telegraph instruments similar to that which is employed in telegraphing; and this movement being frequent and irregular, ordinarily renders it impossible to transmit intelligible signals. During the aurora of September 2, 1859, the currents of electricity on the telegraph wires of the United States were so steady and powerful that, on several lines, the operators succeeded in using them for telegraphic purposes as a substitute for the battery; that is, telegraph messages were transmitted from the auroral influence alone, without the use of any voltage battery. This result clearly proves that the aurora develops on the telegraph wires an electric current similar to that of a violent battery, and differing only in its variable intensity".

A few years later Sophus Tromholt (Chap. 6.5) also discussed this phenomenon, and in his book *Under the Rays of the Aurora Borealis* published in 1885 he writes:

"– on the contrary exceptional Disturbances in the Telegraph Wires occur at the time of great Auroras; since in these Wires, electrical Currents occur, which act against and often completely

Fig. 9.1. This picture shows Eskimos hunting beneath northern light. As one can see, the light emissions give excellent hunting conditions. The similarity between this picture and the Norwegian hunting scene in Fig. 1.1 is striking

cancel these Currents, by the help of which Telegrams are sent from Station to Station. Now and then telegraphing all over the Country can for this Reason be broken and precluded for shorter or longer Times; if one then does not make use of, the Perturbation current itself to send the Despatches, as has sometimes been done".

Thus, Tromholt also called attention to the fact that induced voltages could be so intense in telegraph cables that one could telegraph without having the batteries connected.

During some of his numerous auroral expeditions throughout Norway, Professor Carl Størmer made use of these currents on the telephone wires as an alarm indicating the presence of the northern light.

In his book *The Aurora* (1951) Professor Leiv Harang (Chap. 11.5) describes a few events in which currents induced by the northern lights in telephone wires were so great that the fuses in the circuit were burned out. In particular he describes one incident which occurred on 24th March 1940. The following is an excerpt from this book in which he describes the events at Lødingen:

"At 17 hours all fuses for current (0.5 Amp) on all lines burnt through. Smoke came from burning insulation material in coupling racks. The voltage between Earth and the cables could not be measured as no meters for such high voltages were available.

Fig. 9.2. The curves show disturbances caused by auroral currents which were observed in the power line systems at Kirkenes in Norway. Oscillations with amplitudes as large as 25 Amp are not uncommon

A 230-volt valve (vacuum tube) coupled between Earth and a cable after the most intense phase of the perturbation lighted up very intensely and had to be uncoupled at once in order to avoid damage".

According to Harang there was one instance where voltages were greater than 50–60 V km^{-1}. In practice telephone companies have now installed devices which are supposed to protect the network of telephone lines against destructions such as this, but they are not always successful.

9.3 Disturbances on Power Lines During Periods of High Solar Activity

As in telephone lines used in the auroral zone, large and variable currents can also be induced in long power lines during strong auroral storms. Such disturbances have only been studied to a very limited extent. In Norway, for instance, such studies were not initiated until 1980. Geomagnetically induced currents which flow to the actual ground from the equivalent ground point in the main transformers are measured. With only a very limited amount of data, however, it can be concluded that currents between 1 and 100 Amp often flow due to auroral disturbances, and potential differences corresponding to 1 volt per kilometer on the power line have been measured. During the period when observations have been made no serious breakdown of equipment due to auroral disturbances has occurred in Norway. Similar measurements initiated in North America in the beginning of the 1970's have shown potential differences corresponding to as much as 10 V km^{-1} in the wires and currents up to a few tens of Amperes have been observed in the transformers. Scientists in the USA report instances of transformer breakdowns during periods of high solar activity. We know of no such transformer breakdowns happening until now in Norway. However, it should be noted that this is a problem of very high public importance, and should be investigated thoroughly in order to de-

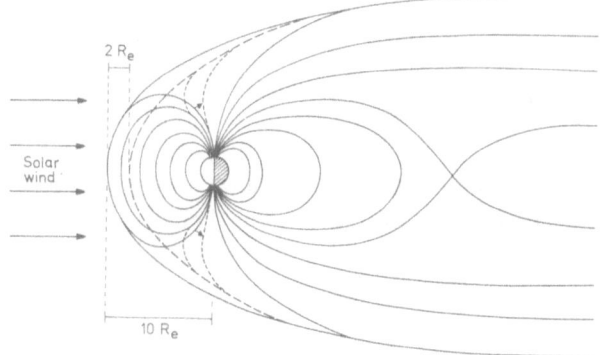

Fig. 9.3. The figure shows a simplified model of the Earth's magnetosphere. Close to the Earth at a distance of about four Earth radii the magnetic field has the shape of a dipole field. Further out it is stretched out into a tail on the nightside and compressed in front on the dayside. The paraboloid-shaped surface encircling the Earth forms the border of the Earth's magnetosphere. During quiet conditions the front of the magnetosphere is situated at a distance of about 10 Earth radii on the dayside whereas during strongly disturbed conditions the front of the magnetosphere can be shifted up to 2 Earth radii closer to the Earth. The *parabolic dashed line* shows this displacement during a hypothetical disturbance; the *dashed lines* emanating from the polar regions illustrate magnetic field lines which will be swept across the pole backwards towards the tail during such disturbances

velop protective devices to prevent such breakdowns from occurring in the future.

9.4 Magnetic Storms and Cable Communication

The most extensive magnetic disturbances are called geomagnetic storms. During these events an enormous number of charged particles propagate through interplanetary space at a speed of approximately 400 m s^{-1} and encircle the Earth. At a distance from the Earth of three–five Earth radii, an intense electrical current is established. The effect of this so-called "ring current" can easily be measured by magnetometers on the ground; reduction in the magnetic field is up to 1%. Intense currents and voltages can be induced in long cables at low latitudes by this "ring current". For instance, during the previously mentioned geomagnetic storm of March 1940,

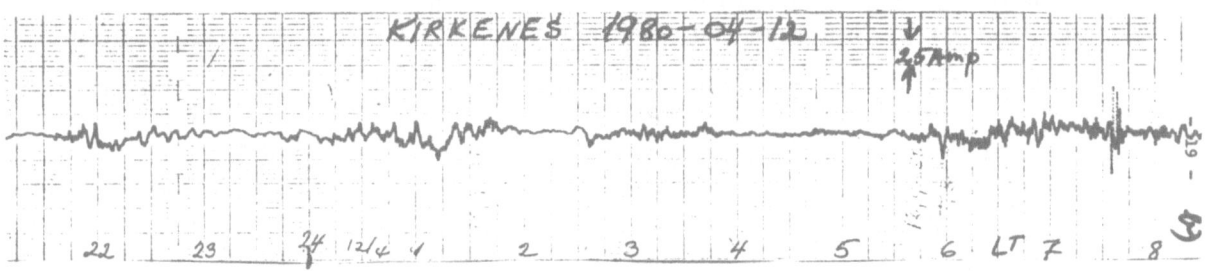

voltages up to 2,650 V were induced in the Atlantic Cable between Newfoundland and Scotland. The storm did not lead to a complete blackout in communications, but it did almost completely distort the conversations from West to East while conversations in the opposite direction were unaltered.

One of the strongest storms in modern times occurred on 4th August 1972 which afforded an opportunity to investigate more closely such cable disturbances. In a 30-min period, a breakdown took place in a cable between Iowa and Illinois, U.S.A. A closer examination of the breakdown revealed that it was caused by induced currents from far outside the Earth. The currents were so vigorous that they shifted the magnetosphere towards the Earth by about one Earth radius on the dayside (Fig. 9.3).

9.5 Effect of Auroral and Geomagnetic Disturbances on Navigation

The Earth is a permanent, gigantic magnet which has a field of geometric form like that of a long bar magnet. The magnetic field is constant except for a few localized and well charted irregularities due to magnetic ore deposits. Even though the Earth magnet changes its position over the years, the change is very gradual and well-known. Thousands of years ago it was discovered that a long, thin magnet freely suspended would always align itself in one direction in the Earth's magnetic field, and in the Northern Hemisphere this direction is quite close to the Pole Star. Such a long, thin, freely suspended magnet refined through the ages and used as a direction finder is known as a magnetic compass.

The magnetic compass is a reliable aid to navigation in areas of the Earth away from the auroral zones. However, within the auroral zones, when there are large solar disturbances, electrically charged particles from the Sun penetrate the Earth's magnetosphere and create strong electrical currents in the Earth's ionosphere at about 120 km altitude. These electrical currents cause large temporary perturbations in the Earth's magnetic field. Such perturbations, called geomagnetic disturbances, turn the

compass needle out of its true reading by an amount dependent upon the strength of the disturbances (Fig. 9.4). A navigator travelling in the polar regions and relying upon a magnetic compass must be aware of these geomagnetic disturbances which usually accompany auroral displays. However, with modern means of communication usually available, other reliable navigation methods exist both for sailing and flying.

In earlier days shipwrecks occurred which were thought to have been caused by compass errors due to disturbances in the Earth's magnetism. One such shipwreck occurred close to Bear Island before the Second World War, but geomagnetic disturbances were never proven to be the basic cause. Some bush pilots in Alaska have claimed that airplane crashes were due to compass errors caused by geomagnetic disturbances.

A very popular sport in Scandinavia today is orientation racing – a cross-country race in which the runners must plot their own course with a map and compass. In this kind of sport it is quite possible that a geomagnetic disturbance such as the one shown in Fig. 9.4 could cause trouble to the participants. The geomagnetic effect of an auroral disturbance can turn a compass needle more than 10° from its equilibrium position within half an hour. Plotting a course by relying on a magnetic compass during such an event would lead the runners astray.

9.6 Geological Survey and Geomagnetic Disturbances

One purpose of a geological survey is to map the location of minerals, metals, oils, gases etc. In this type of survey, magnetic observations are a very important part of field operations. With geomagnetic

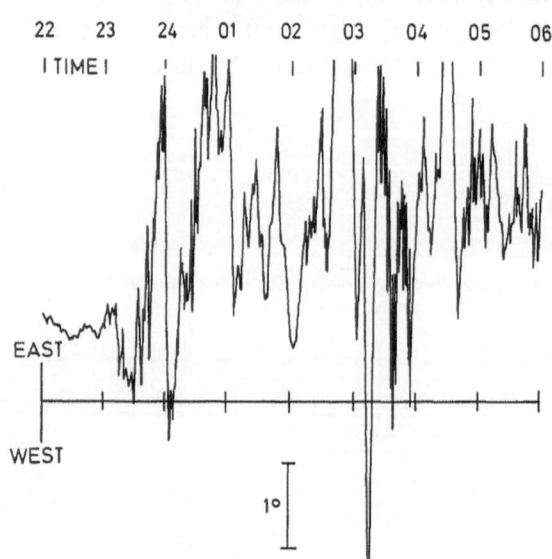

Fig. 9.4. A recording of the displacement of the magnetic needle in Tromsø on May 2–3, 1977, during a large auroral disturbance. The recording shows the variation in the direction of the needle with respect to true north; the corresponding length of 1° variation is also included. Between 0230 and 0330 UT in the morning of May 3rd, the needle shifted by more than 10°

114

measurements one tries to survey the Earth's magnetic field strength in detail and to search for local anomalies due to veins of iron ores etc. The measurements can be accomplished from airplanes or by patrolling the area with portable magnetometers which will reveal variations in the Earth's magnetic field. At high latitudes, as for instance in Scandinavia and Alaska, many of these variations may be due to geomagnetic disturbances occurring during the time of the survey. It is therefore very important when analyzing such data to be aware of such disturbances in order to obtain a high degree of accuracy in the survey. This is especially important in surveying for oil and gas sources on the sea floor. Oil drilling is very expensive and accurate surveys are extremely important. When such surveys are made in the polar region, many holes can be drilled in vain if geomagnetic disturbances are not properly taken into account.

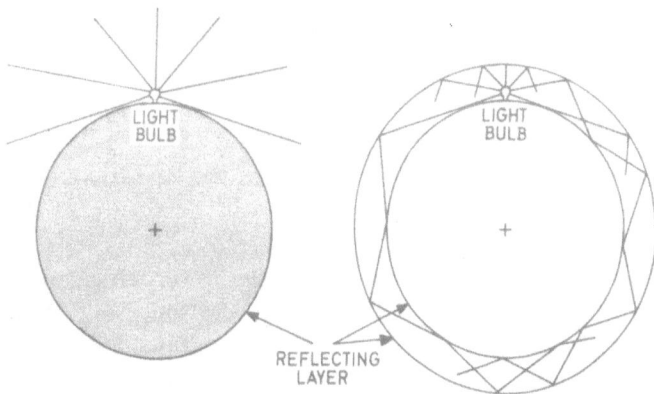

Fig. 9.5. If a small light bulb is placed on the top of a sphere covered by a mirroring layer, only a very small area of the sphere will be illuminated. This area will be determined by the "free horizon" as seen from the position of the bulb, the rest of the sphere will be in the shadow. By putting another sphere covered by a mirroring layer on the inside, outside the first sphere, the latter will be completely illuminated. For propagation of radiowaves in the Earth's atmosphere a similar situation exists. Radiowaves with the proper frequency (1–20 MHz) will be reflected from the ionosphere (100–300 km above ground). The height of the "mirror" determines how many times the wave must be reflected in order to propagate from one position to another; when the ionosphere is low, many reflections are needed between two points; when high up only a few reflections are needed

9.7 Auroral Disturbances and Radio Communication

In the 1930's scientists first began to understand the effects of auroral disturbances and ionospheric currents on radio communications in the polar regions. Under normal or unperturbed conditions, the ionosphere acts like a mirror to radiowaves in the frequency range 1–20 MHz. Radiowaves can propagate beyond the free horizon with one or more reflections between the ionosphere and the ground (Fig. 9.5) since the ground also reflects radiowaves.

Radiowaves pump energy into free electrons along the wave path. In the auroral zone, auroral disturbances are responsible for very many more free electrons (up to one million cm^{-3}) than during quiet times (10,000 or fewer cm^{-3}). Under such disturbed conditions radiowaves will be heavily damped or absorbed (Fig. 9.6). In order to obtain continuous radio communications across the auroral zone and the polar cap it is therefore important to know the optimum wave frequency to be used at any given time. Throughout the world a great effort has been made to improve the frequency forecast service and to ensure the best possible radio communications at any time. Physical relationships, between the source of the disturbances (the Sun) and disturbances in radio communications, are very complicated and only partly understood. This in itself justifies continued intensive research in this field. Secure radio communication is very important in the modern community, both in peace and war.

In the 1960's and 1970's it was thought that by using higher frequencies and avoiding the ionosphere by using line-of-sight radio link stations, problems related to the unstable ionosphere could be avoided. Now, however, it has been realized that such a system is extremely expensive and very vulnerable, which makes it most difficult to maintain during conflict or a war.

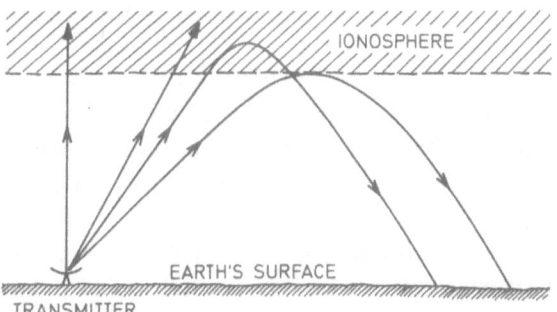

Fig. 9.6. Radiowaves with the same frequency propagating in different directions from the transmitter will be affected differently by the ionosphere. Waves hitting the ionosphere at an angle will be reflected, depending upon the angle; the larger the angle, the higher will be the reflection point. If the angle is larger than a certain critical angle, the wave will penetrate the ionosphere and escape into space. When a strong auroral display exists, an absorbing layer can be built up underneath the "normal" ionosphere. When radiowaves pass through this absorbing layer with a high density of free electrons, some of the energy will be absorbed from the wave and transferred to the electrons such that the wave can be completely annihilated without any chance of reaching the receiver

9.8 Biological Influences on Man
Due to Electromagnetic and Particle Radiations
From the Sun

The Earth's magnetic field shields us very efficiently against particle radiation from the Sun, while the thin ozone (O_3) layer at 20–30 km altitude protects us against solar ultraviolet and X-rays. In the following we will briefly mention biological effects on man due to unusually intense particle and electromagnetic radiations from the Sun.

During disturbed solar conditions particles with extremely high energy are emitted from the Sun. Since the penetration depth into the atmosphere is roughly proportional to the particle energy, the most intense particles may reach down to say 30 km, and in a few cases even below that. Within the ozone layer these particles will dissociate the N_2 (nitrogen) molecules to N (nitrogen) atoms. The N-atoms may react with ozone (O_3) and then form the NO-gas (nitricoxide). The following reaction is then possible:

$$NO + O_3 \rightarrow NO_2 + O_2;$$

i.e., the density of ozone will decrease. The NO_2 gas then reacts with oxygen according to the following formula:

$$NO_2 + O \rightarrow NO + O_2.$$

Thus, this last reaction uses atomic oxygen. As atomic oxygen which is also needed to produce ozone ($O_2 + O \rightarrow O_3$), is a minor constituent at this altitude, the net result is that high energy solar particles reduce the ozone concentration. During intense and long lasting solar activity, the ozone layer may be reduced to such an extent that ultraviolet and X-rays from the Sun penetrate down to the Earth's surface. These radiations are dangerous to human beings.

Biological influences on man by natural and artificial microwaves, radiowaves, electric and magnetic fields, powerline frequencies, etc, have been discussed in the literature during the last two decades, and a lot of controversial results have been published. However, we feel that the question of biological effects on living organisms has not been adequately investigated. It should be borne in mind that these kinds of studies are difficult, because the

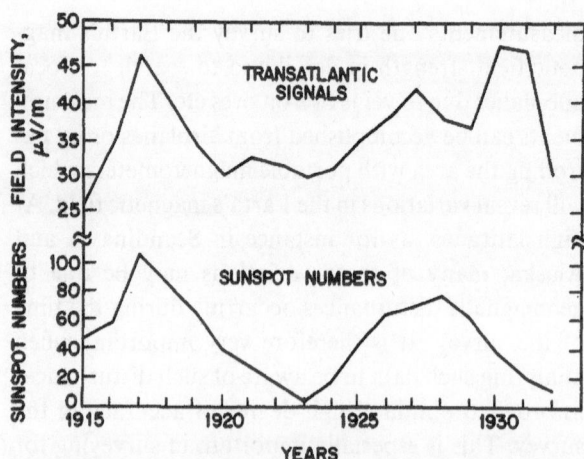

Fig. 9.7. This figure illustrates a relationship between variations in the electric field intensity measured in a transatlantic cable and variations in the sunspot number, thus indicating the important effects of geomagnetic disturbancies on man-made systems (L. J. Lanzerotti, J. Atmos. Terr. Phys. Vol. 41, 787, 1979)

reactions may be due to combined effects, internal and external – very complex problems which cannot be isolated. The effects of ionizing radiation on man will probably receive more attention in the near future.

Some of these effects may be even more serious in the future because the Earth's magnetic field is constantly decreasing. During the last 150 years the geomagnetic field has decreased by 8%. If this trend should continue, the protection effect of the Earth's magnetic field will be markedly less. Thus, high energy particles will penetrate deeper and deeper into our atmosphere.

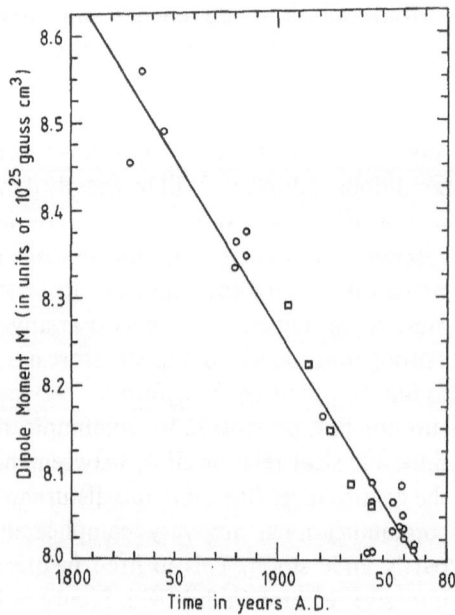

Fig. 9.8. The variation in the strength of the geomagnetic dipole moment (Earth's magnetic field strength) through the last 150 years. An approximate steady decrease has been observed

116

Two very beautiful and realistic lithographs were made by the Danish painter Harald Viggo Greve Moltke (1871–1960) at about the turn of the century

Some original and decorative illustrations made in the 1930's by correspondents of Størmer

The *top* illustrations are very realistic and decorative, the *lower*, however, are more artificial and original

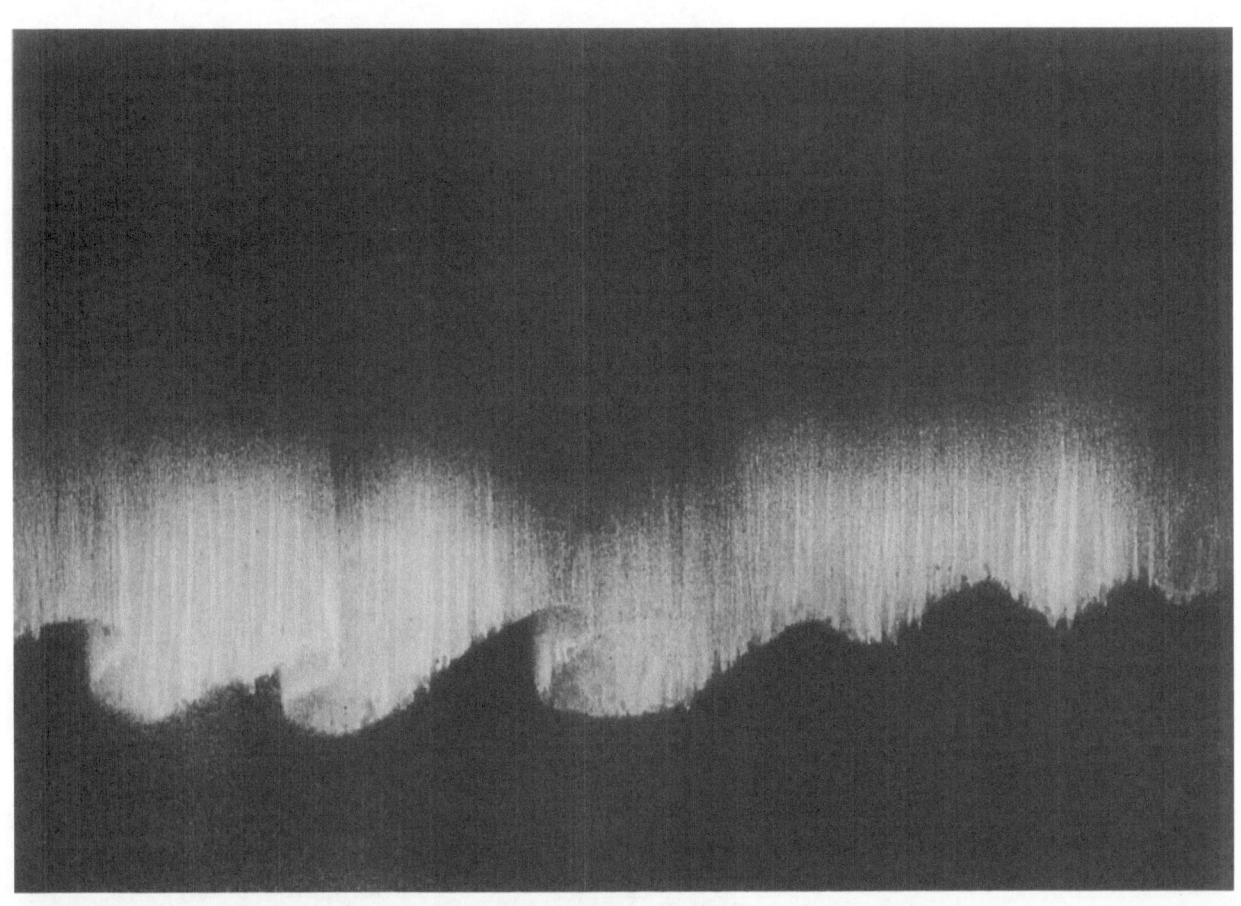

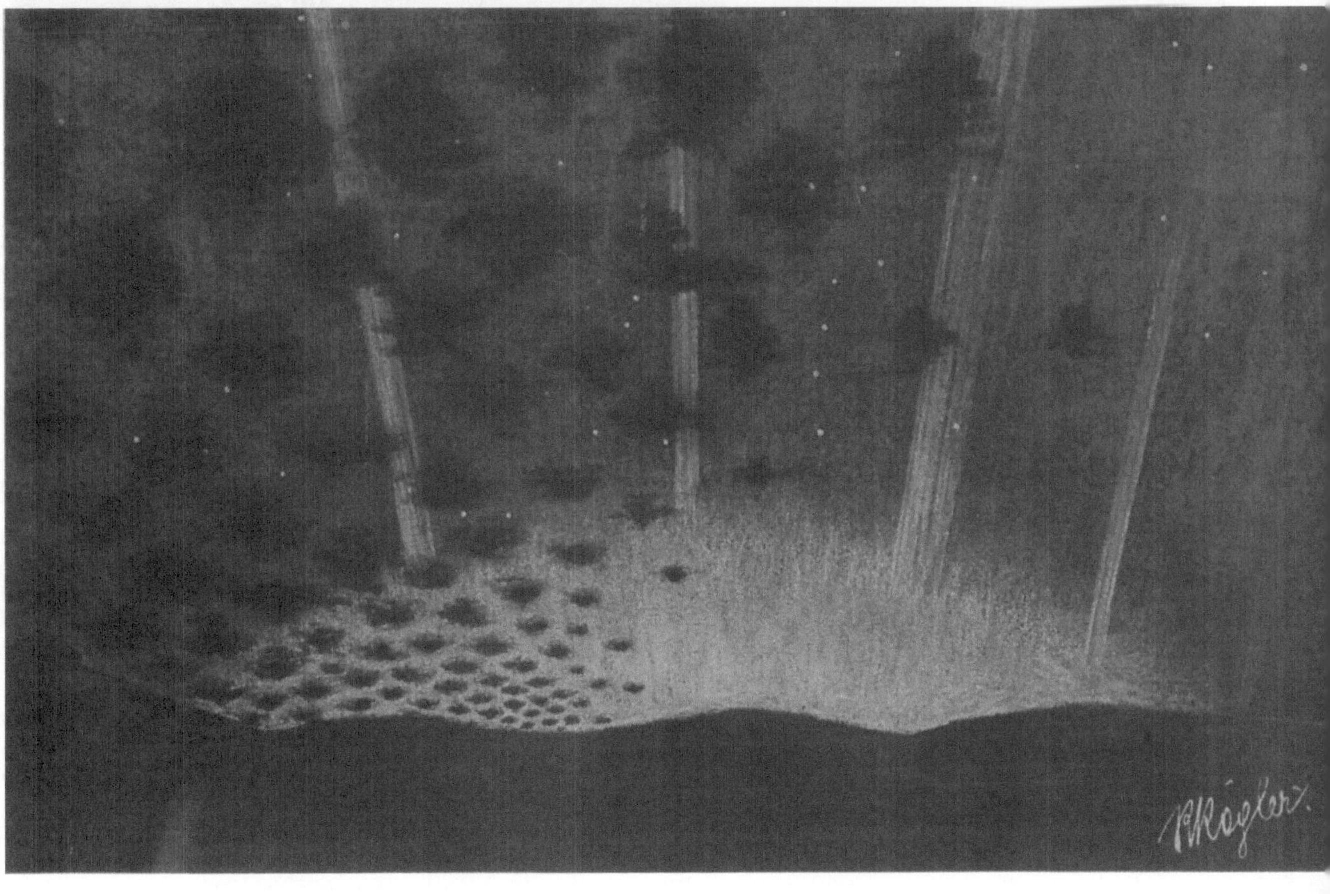

Page 120
The northern light seen at Svalbard January 1979. (Photo K. Henriksen)

Page 121
Top almost identical auroral forms observed simultaneously in the northern and southern hemispheres. The pictures are observed from two aeroplanes at approximately the same geomagnetic field line in the two hemispheres (Courtesy Geophysical Institute, Alaska)

Bottom Auroral drapery (Photo A. Egeland)

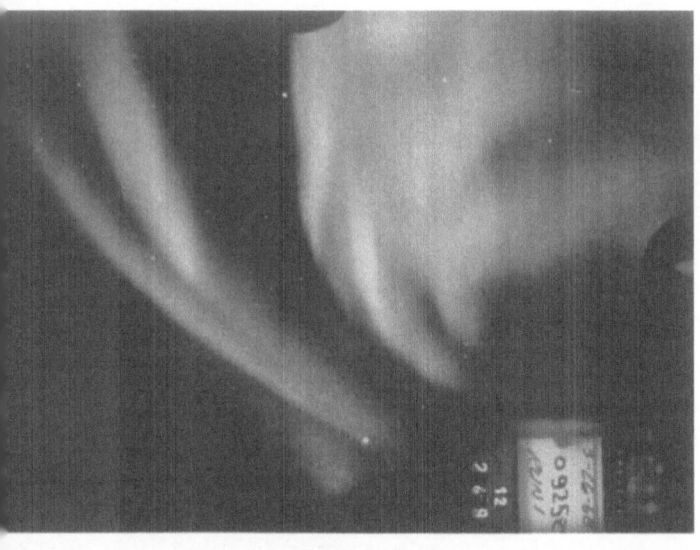
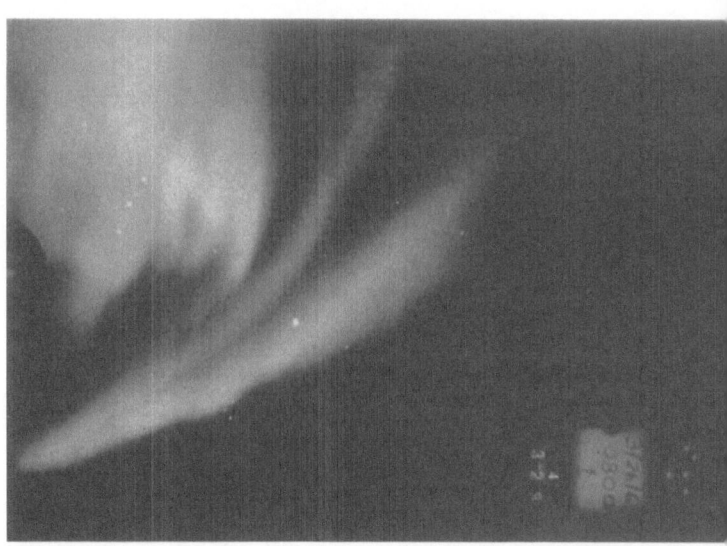

Top. An eruption on the sun.

Middle. An illustration of the position of the auroral oval relative to Scandinavia and Svalbard (Spitzbergen), at different times of day. (*NÅ* Ny-Ålesund; *T* Tromsø; *K* Kiruna; *U* Uppsala). The geomagnetic north pole is marked by +. The colours of the northern light are also illustrated

Bottom. The colour composition of the solar and auroral spectrum

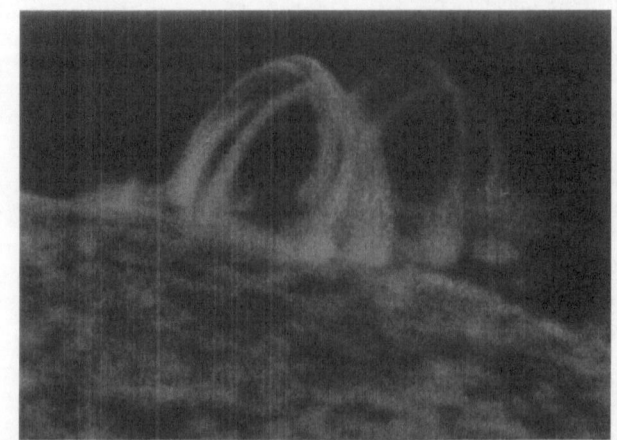

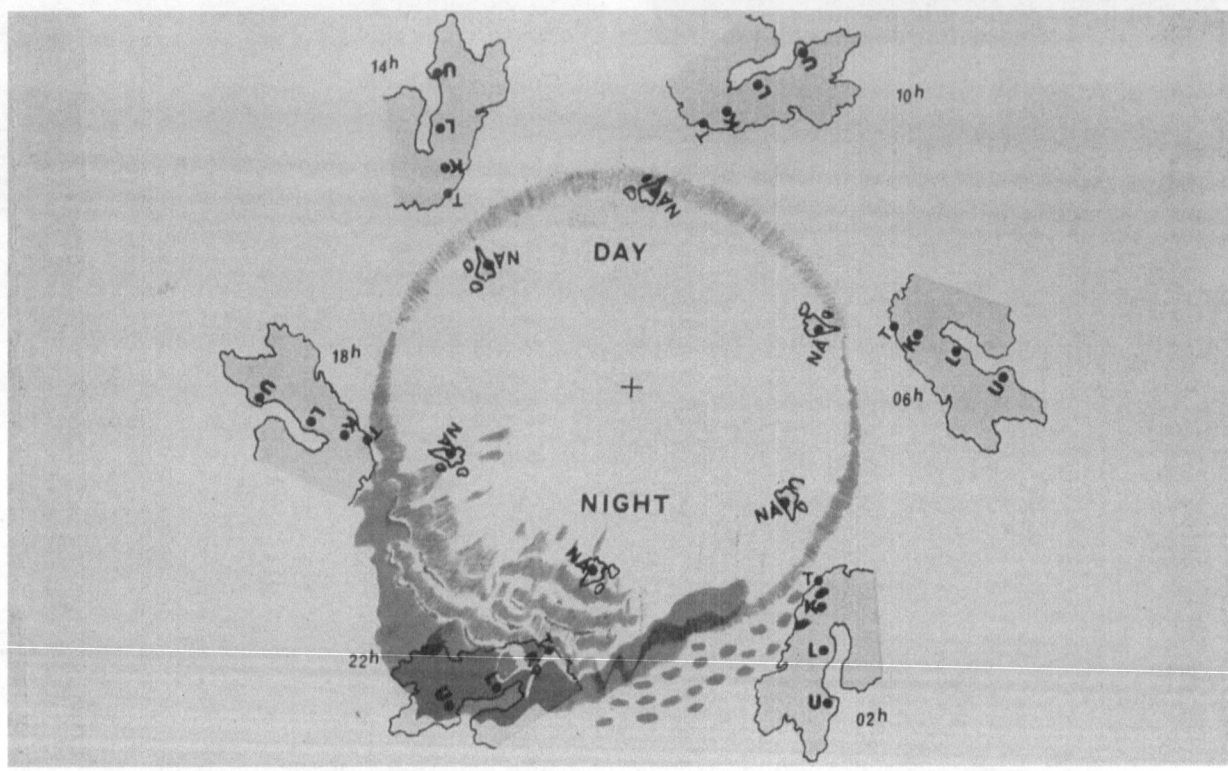

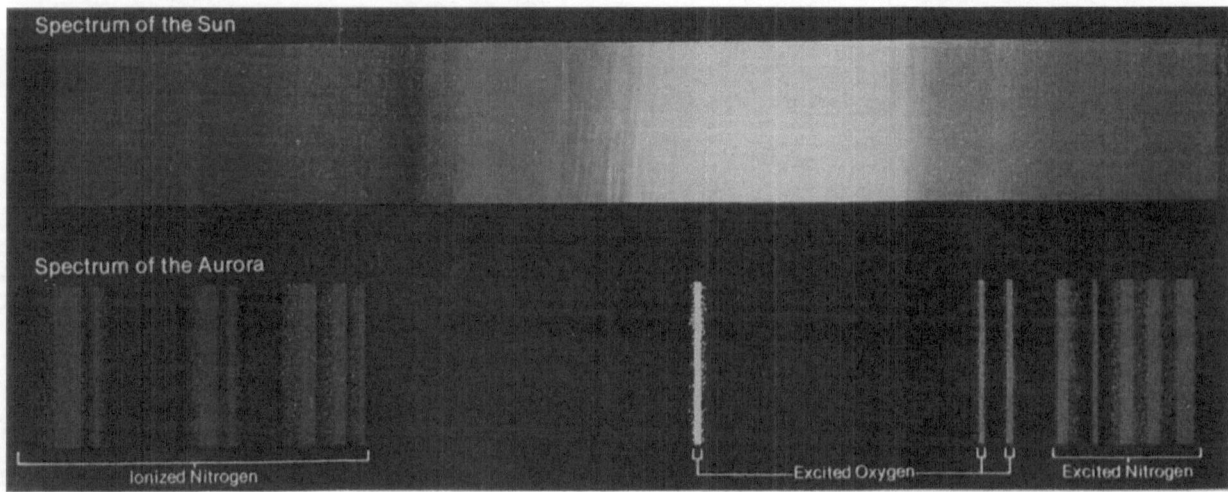

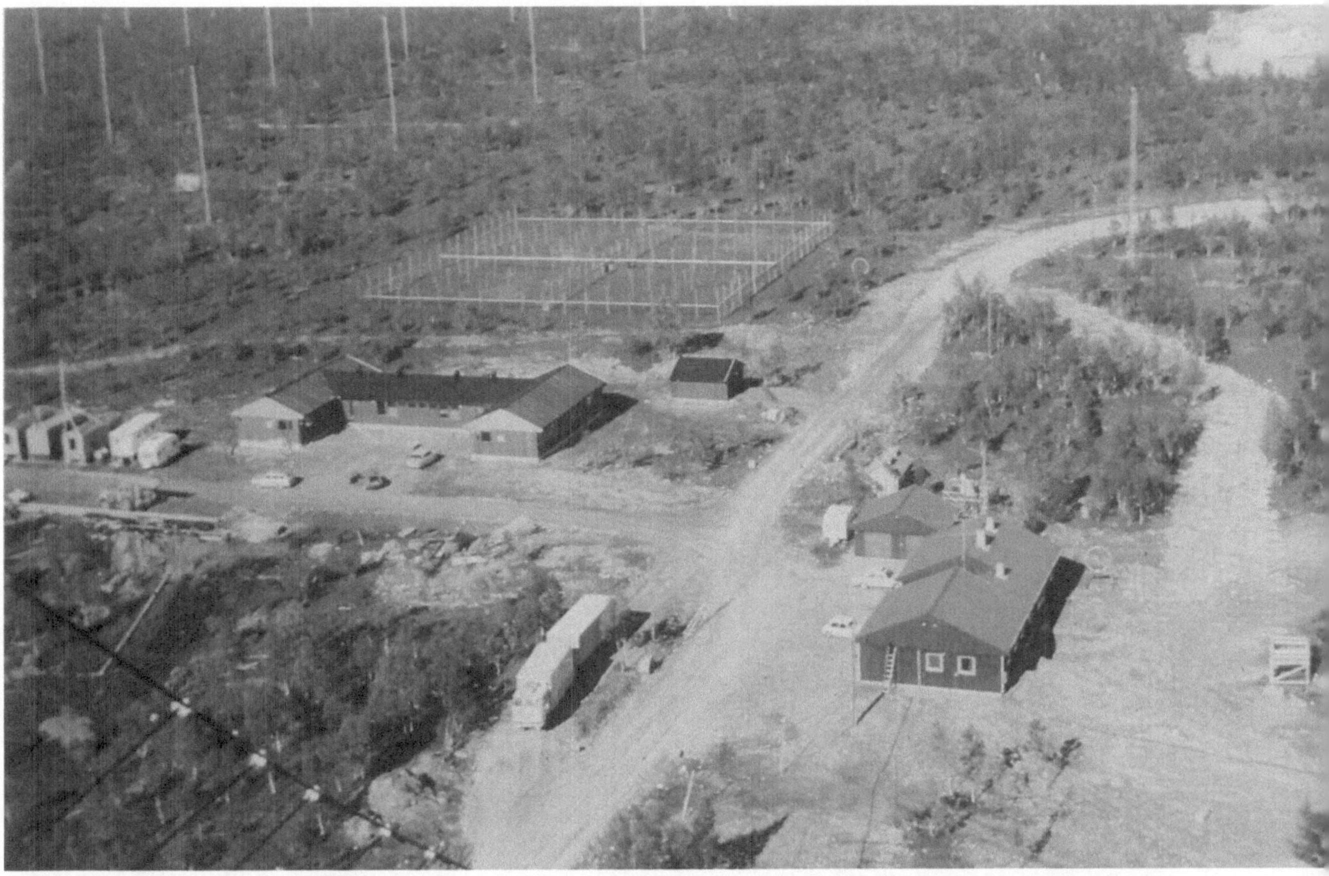

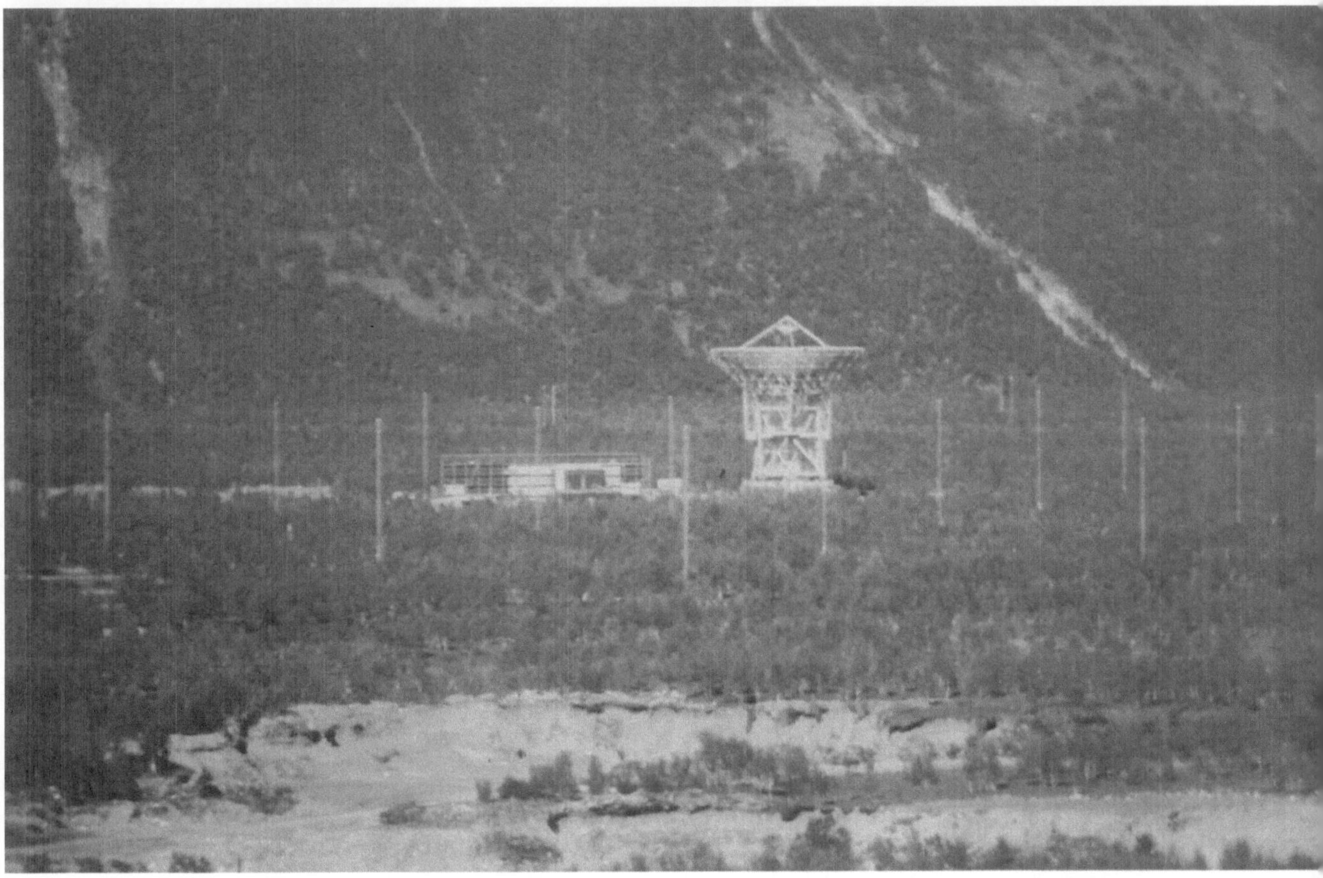

Page 123

Upper part. The University of Tromsø field station to the right and the Norwegian-German Heating Facility to the left, at Ramfjordmoen close to Tromsø (Photo K. Folkestad)

Lower part. The EISCAT Facility at Ramfjordmoen Tromsø (Photo K. Folkestad)

Page 124

Upper part. The artist's concept of the ESRO-I satellite in space

Lower part. A simultaneous launch of two rockets from the Andøya rocket range (Photo J. O. Andersen)

124

10 Auroral Research as a Tool to Study the Upper Atmosphere and Near Space

10.1 The Northern Light was Used to Determine the Height of the Atmosphere

Before the 18th century information about the height of the atmosphere was still very vague, but nevertheless it was one of the most disputed aspects of auroral science up until the end of the last century. It was known that the pressure at ground level was equivalent to a column of water about 10 m high. By assuming that the density of the atmosphere is constant with increasing altitude from ground level, this would indicate that the thickness of the atmosphere would be about 9 km.

During this same period of time, observers knew that meteors were the result of extremely small particles at great speeds being decelerated in the upper atmosphere. Since a meteor could be seen simultaneously from distant observation points, the height of the atmosphere at which the meteor occurred could be roughly determined, and this height was somewhat greater than 9 km. However, since meteors were visible only for a duration of the order of seconds, the accuracy of height measurements was rather poor by this method.

Auroral arcs, on the other hand, are stationary for several minutes and are much more useful for estimating the height of the atmosphere than are meteors. Upper air scientists in the middle of the 18th century realized that if the height of these auroral forms could be measured, then the height of the atmosphere could also be estimated. The results were very surprising and partly contradictory. Before the end of the 18th century approximately 50 estimates of the height of auroral arcs had been obtained which ranged from 200 km up to 1,300 km above the ground (Chap. 7.3). The thickness of the atmosphere was shown to be definitely greater than 9 km and these measurements proved that the volume of the atmosphere around the Earth was much greater than had been previously realized. This knowledge had a great impact on man's conception of his environment.

At the turn of the century, Birkeland introduced a photographic triangulation technique to measure the height distribution of auroral forms. Størmer and his colleagues later used this method with great accuracy and found that most northern lights occur at 100–120 km above ground. With modern photographic techniques northern lights as low as 90 km and as high as 1,000 km have been observed.

At the beginning of our century the density and composition of the atmosphere at great heights remained puzzling, but a possible solution to this problem was to be found through auroral research. In 1867 Ångström (Chap. 6.2) had measured the first spectra of auroral light, and Lemström (Chap. 6.3) had shown that these spectra were very similar to emission spectra observed from rarefied air in the laboratory. Identification of the green line measured at 5,560 Å by Ångström was still not possible since it could not be matched with any known line from any of the elements in air at ground level.

When in an excited state, each of the elements emits characteristic lines in its spectrum which are in such an ordered array that each element present can be determined from these spectra. With such knowledge, spectroscopic studies of the northern light made it possible to identify the gases composing the upper atmosphere. Auroral spectra could also be used to make temperature measurements of the atmosphere at auroral heights.

In 1923, two Canadian scientists, Sir John McLennon and G.M. Shrum showed that the auroral green line was due to a metastable transition of atomic oxygen. The transition is not absolutely forbidden but rather the statistical probability for the transition to occur is very small, so that the atom remains in its excited state for a long time before radiating. In fact, since the atom can remain in its excited state for almost a second before emitting the green light, it can lose its excess energy by collision with other gas particles and no light emission is observed. This was mainly why it was so difficult to identify the emission in the laboratory. Due to the difficulty, in those days, of obtaining a vacuum in the laboratory comparable to that at auroral altitudes, excited oxygen atoms in this forbidden state would collide with other gas particles or the walls of the vacuum chamber before emitting light. Consequently, no emissions at 5,577 Å were observed from artificial sources in the laboratory until 1923. 5,577 Å is the exact wavelength of the auroral green line identified by Ångström.

10.2 Composition of the Atmosphere in the Height Region 100–300 km Above Ground

Density and composition of the atmosphere together with the energy of the precipitating particles determine the colour of auroral displays. Therefore the height and colour of the northern light reveals a wealth of information on the composition of the upper atmosphere.

Partly for physical and partly for practical reasons, airspace around the Earth has been divided into several approximately spherical shells which are different in terms of temperature, density, composition and Motion. The demarcation lines between these spheres are not sharp and are therefore called pauses.

Up to a height of 60–70 km the Earth's atmosphere is homogeneous in composition and electrically neutral. Above a height of 70 km free electrons and ions appear and this region is called the ionosphere. Charged electrons and ions interact with radio waves making the ionosphere very important in long distance radio wave communication and navigation. Only a tiny fraction of atmospheric gases at auroral heights is ionized; the major part is still neutral.

The temperature of the neutral atmosphere increases markedly between 100 and 200 km. This temperature increase is caused by heating due to absorption of solar radiation. About 40% of the absorption occurs in the upper atmosphere which results in the important process of atmospheric molecules being dissociated into atoms.

At wavelengths below 2,400 Å, solar radiation splits oxygen molecules (O_2) into oxygen atoms. Nitrogen molecules require a higher energy to be dissociated and solar radiation at wavelengths less than 1,000 Å is necessary for this process to occur. Dissociation of oxygen molecules begins at about 70–80 km altitude, and at about 95 km there are almost equal numbers of oxygen atoms and molecules. At 150 km altitude there are about ten times as many oxygen atoms as molecules. The dissociation of nitrogen molecules (N_2) takes place mainly at altitudes between 120 and 200 km. Oxygen, which is lighter than nitrogen, tends to "float up" and at about 200 km altitude there are approximately equal amounts of oxygen and nitrogen. The dominant gaseous species between 100 and 300 km altitudes are atomic oxygen and nitrogen, whereas at ground level molecular nitrogen is by far the dominant species. The composition of the atmosphere therefore changes considerably with height and at the upper altitudes it changes with the solar zenith angle and with the time of day and the season.

Until about 1910 it was thought that the atmosphere above 100 km altitude consisted primarily of hydrogen and helium. Spectroscopic studies of the northern light at the beginning of this century, more than any other research, contributed to a thoroughgoing analysis of gaseous constituents above 100 km.

Both pressure and density of the atmosphere decreases rapidly with altitude. Molecular density of the air is about 10^{19} cm^{-3} at ground level and decreases to approximately 10^{13} and 10^7 molecules cm^{-3} at 100 and 300 km altitudes, respectively. Similarly, atmospheric pressure decreases from 760 mm Hg at ground level to about 10^{-4}, 4×10^{-6} and 10^{-6} mm Hg at 100, 150 and 200 km altitudes, respectively.

One can illustrate the amount of air surrounding the Earth by the following example: If the Earth's atmosphere were compressed in such a way that the density and temperature were of constant values and equal to those at ground level, then the thickness of the atmosphere would be only approximately 8 km or 1/800th of the Earth's radius. There is no clear demarcation of the outer boundary of the Earth's atmosphere, but compared to the dimensions of the Earth, the thickness of the atmosphere is somewhat analogous to an apple skin compared to the size of the apple itself.

Farther out from the Earth we find the lighter gases such as hydrogen and helium are predominant in the atmosphere at very low densities, and the greater number of these gases are ionized.

10.3 The Solar Wind – The Main Source of the Northern Lights

Before the advent of the space age, auroral scientists were quite limited in the scope of their endeavours

126

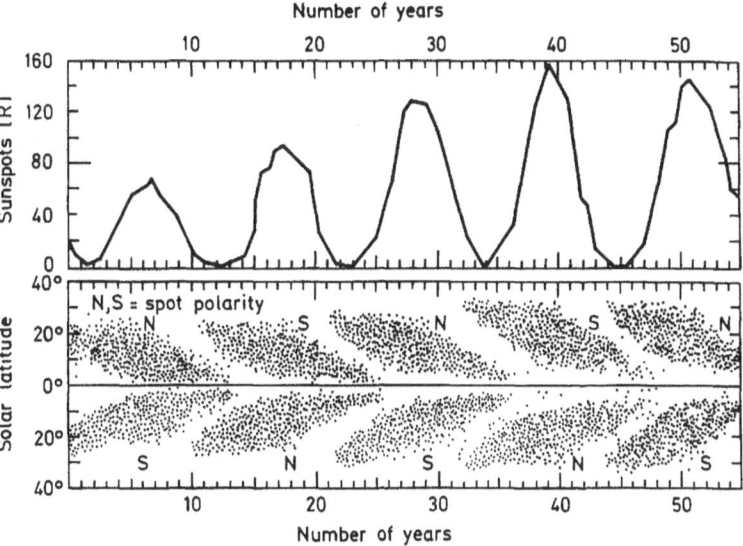

Fig. 10.1. *The upper panel* shows the approximate variations in the sunspot number over a period of about 55 years. The distance in time between the maxima – one solar cycle – is approximately 11 years. *The lower panel* shows approximately how sunspots are distributed on the solar surface during the same period as in the upper panel. At sunspot minimum new sunspots are created at high heliocentric latitudes, the spots occur progressively more equatorward as the solar maximum approaches. Also notice that the magnetic polarity changes from north (N) to south (S) in consecutive cycles

and spent much of their effort in describing the morphology of the northern light from ground-based observations. At the turn of the century, knowledge of the source mechanisms creating this light was very vague. Kristian Birkeland's elegant terella experiment had strongly suggested that the northern light was the result of an interaction between charged particles of solar origin and the Earth's magnetic field, but not all auroral scientists were totally convinced by his experiment. The first rockets which were launched into auroral arcs and the satellites which were put into orbit in the 1950's measured energetic particles penetrating the Earth's upper atmosphere in the auroral zones. With these data even the most sceptical opponents of Birkeland's theory had to back down.

An instructive way of understanding how a northern light is produced is to think of the atmosphere as a gigantic TV-screen which brightens when it is bombarded by an energetic stream of particles. By analyzing the light emitted by the screen one can obtain information about the energy of the bombarding particles and the composition of the screen or the atmosphere. Similar analyses of auroral light can also help in determining what happens in the atmosphere regarding interactions between the neutral gas and solar particles.

Since the basic source of energy for the northern light is the Sun, the occurrence of the northern light and its strength is determined by the status of the Sun. The Sun is situated in the Orion arm of the Milky Way, and the distance from the Earth to the Sun is approximately 150 million kilometers. Since light propagates at a speed of 300,000 $km\,s^{-1}$ it takes approximately 500 s or a little more than 8 min for sunlight to reach the Earth.

The temperature at the solar center is approximately 15 million degrees, and at such extreme temperatures energy is released by so-called nuclear fusion reactions. Every second of time the Sun radiates approximately 10^{27} W omnidirectionally, and even though the Earth receives only one part in 200 million of this radiation, the Earth on the average still receives the equivalent of 1.25 $kW\,m^{-2}$.

Temperature in the solar atmosphere (solar corona) is equivalent to a few million degrees. At such high temperatures the resulting pressure forces are much larger than the attraction forces due to the Sun's gravity, with the result that some of the gases in the solar atmosphere escape from the Sun. At the very high temperature of the solar corona these gases will be ionized and they consist of electrons and ions (protons) with negative and positive charges, respectively. On the average the flux of negative and positive charges from the Sun will be equal, so the outflowing gas will appear neutral. This stream of particles from the Sun is called the Solar Wind. The particles' density in this wind is approximately 10 cm^{-3}, but this number will vary depending on the activity of the Sun and quite often in such a manner that density will increase with decreasing solar activity. Only a small number of the particles streaming out from the Sun, however, will hit the Earth's atmosphere.

Satellites equipped with sensitive instruments have convincingly demonstrated that this stream of matter escapes from the Sun with an average speed of 400 $km\,s^{-1}$, and therefore takes approximately 5 days to reach the Earth. However, during very dis-

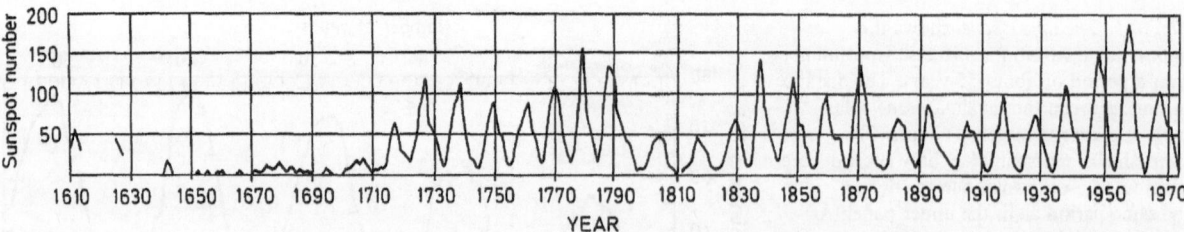

Fig. 10.2. A modern graph of the annual mean sunspot number from 1610 to 1970 showing clearly an 11-year periodicity. Note, however, the very low activity prior to 1715 (The Maunder Minimum) and also that the number of sunspots have varied considerably during the last centuries

turbed periods on the Sun when highly energetic particles are produced, this travel time can be as small as two days.

For a long time it has been well known that the occurrence of the northern light is strongly related to the number of sunspots. These sunspots can have a lifetime of from one to several months (Fig. 10.1) and since the Sun rotates on its own axis in a period of approximately 25 days, or about 14.4 degrees day^{-1} and the Earth in its orbit around the Sun moves at an angular speed of approximately one degree per day, the same area of the Sun will point towards the

Earth every 27th day. This is the reason why similar auroral displays can reappear every 27, and often every 54 days.

In 1843 the German pharmacist and amateur astronomer, Heinrich Schwabe (1789–1875), after many years of systematic observations discovered the so-called 11-year solar cycle (see Fig. 10.2). His result stimulated several attempts at correlating the

Fig. 10.3. Upper panel The annual auroral occurrence frequency in North America from 1775 to 1870 as obtained from data collected by Elias Loomis.
Middle panel Annual auroral occurrence frequency in Sweden from 1720 to 1875 as obtained from data collected by Robert Rubenson.
Lower panel Annual auroral occurrence frequency in Scandinavia from 1720 to 1880 as obtained from data collected by Sophus Tromholt

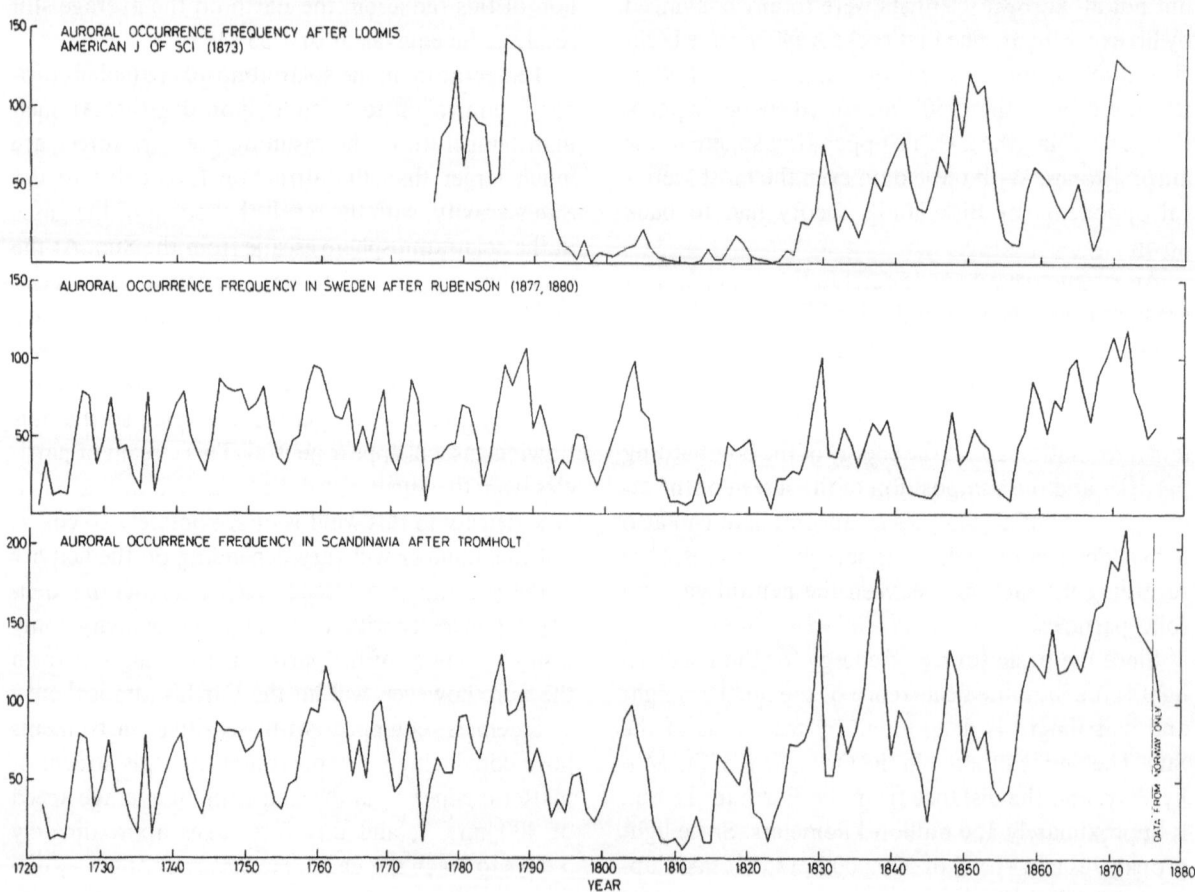

occurrence of the northern light with the solar cycle. In 1870 Elias Loomis was able to show such a correlation which was further confirmed by Rubenson in 1877, Fritz in 1881, and Tromholt in 1902 (published after his death) (see Figs. 2.8 and 10.3).

The physical relationship between the presence of sunspots and the occurrence of the northern light has not been well determined. It appears, however, that the sunspots themselves are not the direct source of the auroral particles but rather that the occurrence of sunspots is related to some active areas on the Sun, often called M-regions which are now thought to be the so-called coronal holes. From these coronal holes high speed particles appear to propagate outward from the Sun.

During special events called solar flares an enormous amount of energy is released from the Sun in the form of high speed particles and waves. Under these conditions, on clear winter nights the northern light can often be seen in Oslo and sometimes even further south.

10.4 The Northern Light Can be Used to Study Properties of the Sun

The primary energy source of the northern light is electrons and protons emanating from the Sun, and therefore the occurrence and intensity of the northern light are determined by solar conditions. Systematic observations of the Sun began about 1830, but since the time when Galileo Galilei (1564–1642) made his first telescope in 1610 and looked at the Sun, several good observations of sunspots have been recorded.

Before 1600 A.D., however, there were sparse sunspot data since the observations were based on sightings with the naked eye. In order to observe sunspots with the naked eye, the Sun has to be partly shadowed by moisture or dust in the air, and such conditions are more likely to occur on islands far out in the open sea or in desert areas. Records of naked eye sunspot observations are available from Japan, China, and Korea. The data available are, of course, very limited, but Fig. 10.4 shows a 20-year average of such sightings from the beginning of the Christian era up until about 1600 A.D. Note that there is a period between 500 and 1100 A.D. when there were very few sunspots seen in Eastern Asia, and this coincides with data from the Viking era in Scandinavia, as mentioned in Chap. 2.

The relationship between the solar cycle and the northern light, however, also makes it possible to use ancient auroral observations to partly reconstruct the solar cycle which existed previous to the observations of Galilei. In particular, a special type of red coloured northern light tends to reappear close to the maximum of the solar cycle, which lends credence to this kind of backlogging of the sunspot cycle. Such red northern lights have an overwhelming appearance and terrified people, particularly during the Middle Ages in Europe. Therefore, one will often find notes in the old annals in Europe of such threatening lights in the sky. In Fig. 10.5 the occurrence of such red northern lights occurring between 1750 and 1940 in Norway has been plotted, together with the annual sunspot number to show the tendency for them, to occur during the maximum phase of the solar cycle.

There are also other methods to study solar activity in historic times. One of these, the so-called C^{14} method, is based on the radioactive isotope of carbon. Also modern research on an oxygen isotope found in the great glaciers can throw light on the behaviour of sunspots.

In addition to the great lull in solar activity in early medieval times there was also a period of great activity between 1,100 and 1,300 A.D. which coincided

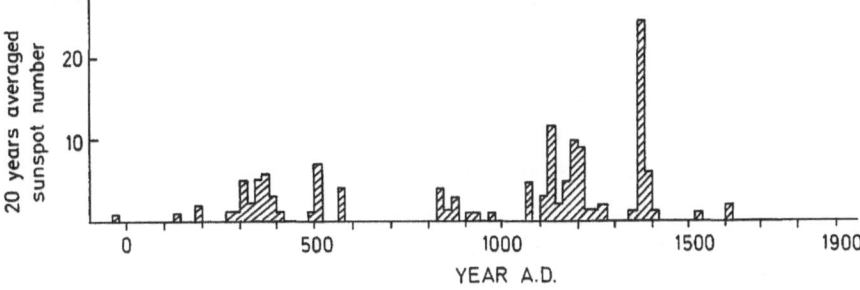

Fig. 10.4. A 20-year average of the number of sunspots observed by the naked eye in China and Korea until about 1600 A.D. Between 600 and 800 A.D. solar activity was very low, while the number of sunspots have a broad maximum between 1100 and 1300 A.D. and a very sharp maximum close to 1400 A.D.

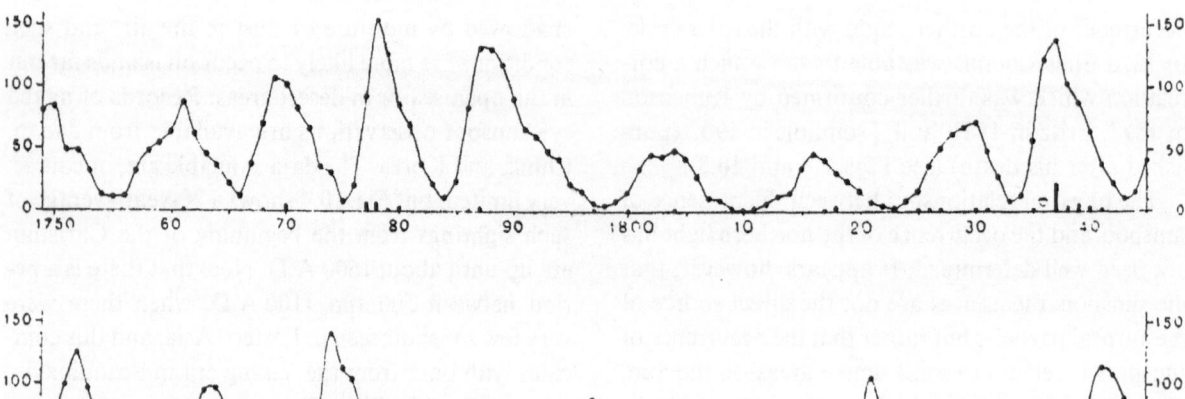

Fig. 10.5. Annual occurrence number of red northern lights over Norway from 1750–1940 shown as dark streaks and compared with the annual sunspot curve according to Størmer

with the time that *The King's Mirror* (Chap. 2.2) was written in Norway. In spite of this, as we have seen the author discusses the northern light in terms suggesting that he himself never saw it and refers to it as something unique to Greenland. This makes it probable that the magnetic pole – and thereby the whole auroral zone – must have changed its position during the last 1,000 years.

10.5 The Magnetic Field of the Earth is the Guideline of the Northern Light

The magnetic compass is a very useful tool for navigation because the Earth itself is a great magnet. The morphology of the Earth's magnetic field at its surface is very similar to the field one would get if the Earth were a gigantic magnetized sphere, or if a huge magnet were placed almost parallel to the Earth's rotation axis at its center. The rotation axis of the Earth is displaced about 11° with respect to the approximate symmetry axis of the magnetic field and because of this one magnetic pole points approximately towards Thule, Greenland. The magnetic field is also approximately symmetric about the magnetic equator which is situated half way between the two magnetic poles. The Earth's magnetic field is twice as strong at the poles as at the equator. At the poles it is about 0.6 Gauss or 0.6×10^{-4} Tesla[18]. The field intensity decreases very rapidly with distance

from the centre of the Earth. At a distance of approximately 6,500 km above the ground the field strength is only one-eight of that at the Earth's surface, while at auroral heights it is 95% of the field strength at the Earth's surface. One can get a very good idea of the direction of the field at auroral heights by noting the alignment of auroral rays (Figs. 3.1 and 7.3). These thin rays which follow the field direction make the "fieldlines" visible.

One can say that the Earth's magnetic field is the guideline of the auroral particles. In reality it is the geomagnetic field lines reaching out into space which guide the solar-wind particles toward the auroral zones. Disturbances on the Sun also lead to disturbances in the Earth's magnetic field, and such geomagnetic disturbances are therefore of great interest in relation to auroral intensity and geographic position.

10.6 The Geographic Position of the Northern Light has Changed with Time

It is well known that changes in deviation of the compass needle from true north have occurred from year to year because the Earth's magnetic field is changing its position. In Oslo, for instance, the deviation of the magnetic needle has changed by 10° during the last century. This variation in the geomagnetic field indicates that it is only partly correct to say that the Earth is a permanent magnet. The reason for these variations in the Earth's magnetic field is believed to

---

[18] Unit of the magnetic field

130

have two different sources; one internal and one external. The internal sources are associated with very complicated electric currents in the interior of the Earth. The effect of these currents is that the field drifts with respect to the rotation axes of the Earth; at present the drift is about 0.25° westward year^{-1}. Another effect is a gradual decrease in the field strength of about 0.07% year^{-1} (see Fig. 9.8). This may appear small and inconsequential, but over a long period of time it can become very important. If, for instance, the decrease of the field strength continues at this rate for the next 1,500 years, the field will be reduced to about 35% of its present value and this would create a serious threat to all life on Earth.

The oldest reliable and accurate observations of the Earth's magnetic field date back only about 200 years. These observations were good enough to determine the position of the magnetic poles and during this time-period, the center of the auroral oval has moved by more than 800 km. Today the north geomagnetic pole is close to Thule, Greenland.

Currently there is a reawakened interest in the behaviour of the geomagnetic field as it existed both in historic and in prehistoric time.

The orientation of magnetized minerals in solidified lava can be used to study how the geomagnetic field was oriented in prehistoric times. In molten lava, magnetized minerals aligne themselves in accordance with the geomagnetic field existing at the place where the lava is petrified.

To determine the orientation of the geomagnetic field in historic times, burned clay and other construction materials can be analyzed. When the material is in a state similar to concrete, the magnetic field will also be "frozen in". The orientation of the geomagnetic field at the time the clay solidified will therefore be preserved in the material. Old construction material and burned clay can therefore reveal important information about the orientation of the Earth's magnetic field, as it existed for example in the Viking era.

Based on such archaeomagnetic studies the position of the north geomagnetic pole has been derived for the years 700 A.D. to 1900 A.D. (see Fig. 2.10) and the result of this research has already been presented for the years 1200 and 1970 A.D. in Fig. 2.11.

10.7 A Simplified Modern Theory of the Northern Light

Most people are familiar with the fact that a loud bang occurs in the surrounding air when an airplane attains a speed which is higher than the speed of sound. A similar effect can be observed when a gas, at supersonic speed, streams past a sphere at rest as is illustrated in Fig. 10.6. (The airstream in this figure

Fig. 10.6. A simplified illustration of the shock front forming around a sphere in a supersonic air stream

Fig. 10.7. A simplified model of the influence of the solar wind on the Earth's magnetic field. The Earth is observed from a point in the equatorial plane at local time 18. The solar wind flows from the left with supersonic (super-Alfvénic) speed, and forms a shock front on the dayside of the Earth. This shock front also is drawn over to the nightside and forms a paraboloid downstream of the Solar Wind. The turbulent region just inside the shock front is called the magnetolayer, and inside this layer is the magnetosphere. The dark areas inside the magnetosphere are the van Allen Zones (belts) which form in the shape of a doughnut around the Earth

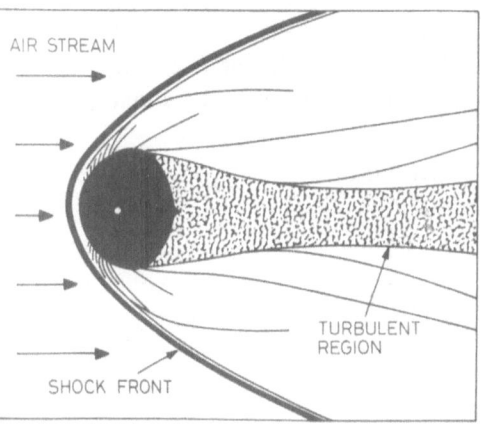

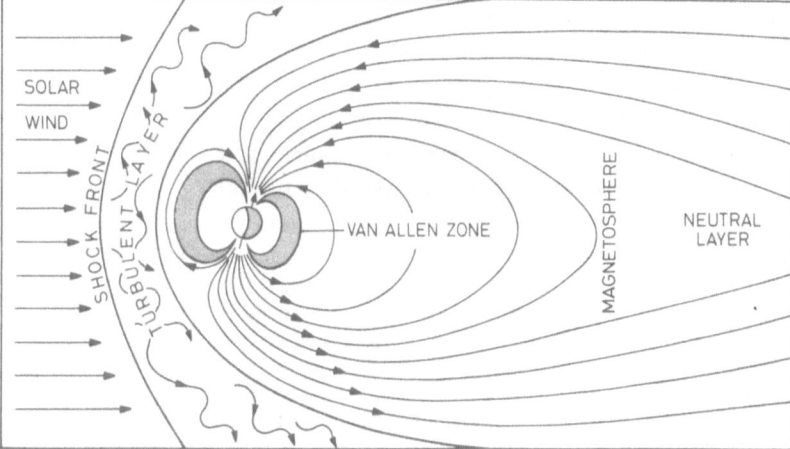

Fig. 10.6   Fig. 10.7

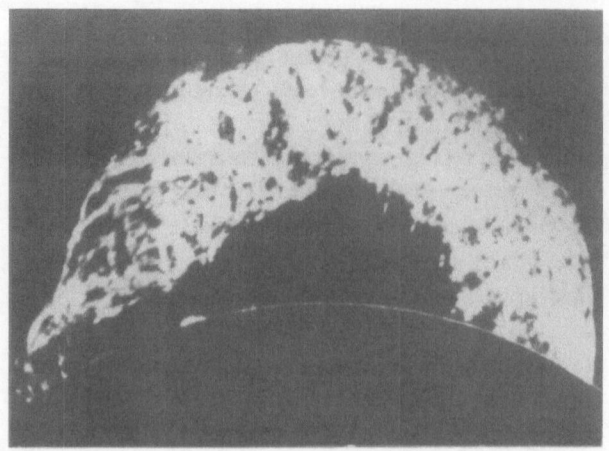

Fig. 10.8. A gigantic eruption on the Sun where dense plasma clouds are ejected into interplanetary space. The solar surface is shown as a narrow light arc at the bottom of the picture

has a speed approximately nine times the speed of sound in the air around the sphere.) A shock front occurs at a small distance in front of the sphere.

When the continuously blowing Solar Wind (Chap. 10.3) reaches the Earth's magnetic field, a shock front occurs in front of the Earth at a distance of approximately 10 Earth radii, during normal solar conditions. This shock front is stretched out in the shape of a paraboloid around to the nightside of the Earth in a direction parallel to the solar wind.

This interaction between the solar wind and the Earth's magnetic field results in a deformed field which is compressed on the dayside and stretched into a tail on the nightside, reaching out as far as the Moon and sometimes even further (Fig. 10.7).

The space inside this paraboloid is called "near space" or the magnetosphere. The magnetosphere is that part of the Earth's atmosphere where the motion of atmospheric gases is controlled by the Earth's magnetic field. Close to the Earth it is limited to the top of the ionosphere, which is about 200 km above ground.

In the tail of the magnetosphere the magnetic field will be oriented almost parallel to the solar wind in such a way that it does not form complete loops but rather becomes open and has only one end anchored to the Earth. The magnetic field lines closer to the Earth have both ends attached to the Earth similar to those of a true dipole field. In the polar regions, however, the field lines are not closed but actually are directed outward into space where they merge with the interplanetary magnetic field lines.

The solar wind in the solar atmosphere, like an ordinary wind, is filled with gusts and gales. Sometimes when there is a strong eruption on the Sun, large clouds of gas are thrown out at great speed from the solar surface (Fig. 10.8). These gas clouds strike the Earth's magnetic field with a vigorous impulse, and the stable situation which may have existed for several

days can be so quickly and strongly disturbed that several days can pass before equilibrium again is restored. Such gigantic encounters between the solar wind and the Earth's magnetic field are very complicated phenomena, and the cause, as well as the effect, is only partly understood. The most spectacular result of these encounters is the aurora or northern light.

The magnetosphere is partly controlled by the Earth's magnetic field. During strong solar disturbances, however, the Solar Wind tends to pull the magnetosphere along with it. These interactions create particle motion in the magnetosphere as the Earth rotates, with her magnetic field around its rotation axis, resulting in electrical currents which float along the magnetic field lines between the poles and the equatorial plane. These currents, called Birkeland currents, complete their circuits through currents vertical to the field lines in the equatorial plane and horizontal currents in the ionosphere at auroral latitudes. A simplified model of this current system is illustrated in Fig. 10.9.

Currents floating around the Earth in near space are the main reason why the magnetic field at great distances from the Earth is so different from the magnetic field of a gigantic dipole. One notices that

Fig. 10.9. A model of the current loops between the magnetosphere and the ionosphere (atmosphere). The currents which have their sources partly in the magnetosphere and partly in the ionosphere are connected in the ionosphere by horizontal currents and in the magnetosphere via a radial current in the equatorial plane

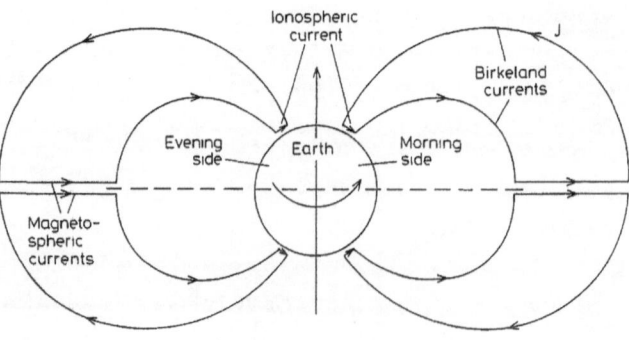

all field lines which are stretched out in the tail have their footpoints close to the magnetic poles. Therefore, the polar regions are less screened from the Solar Wind than regions at lower latitudes. In fact, the ionosphere in the polar regions is coupled to the Solar Wind via electric and magnetic fields from interplanetary space. Ionized gases move across the polar regions in the same region as the Solar Wind. The gases, however, cannot be piled up anywhere on the nightside of the Earth, and therefore a continuity in flow is obtained by a return flow towards the dayside at auroral latitudes. This convection always takes place and is a fundamental dynamical property of the high latitude upper atmosphere and ionosphere. During very active periods on the Sun the velocity of the Solar Wind can be greatly enhanced and the high latitude ionosphere will also increase its speed in such a way that greater currents flow in the upper atmosphere, which can be observed as large fluctuations in the magnetic field at ground level.

It is still not quite clear from where the penetrating particles, which impinge on the high latitude upper atmosphere (and create the northern light), gain their energy. In earlier days it was thought that particles entered the atmosphere directly from the Solar Wind, but observations by rockets and satellites have shown that the particles gain an excess energy on their journey from the Solar Wind to the auroral ionosphere. Satellite measurements have shown that there are energetic particles in the so-called neutral layer in the tail. This layer is present in the central part of the tail region where the magnetic field lines are approximately antiparallel.

Simultaneous observations on the ground and from satellites have revealed that when the typical evening auroral arc moves toward the equator, the plasma (or ionized gas) in the neutral layer approaches the Earth. At the time of an auroral breakup, when the arc loses its form, gas pressure decreases in the neutral layer far back in the tail. While auroral arcs move poleward, plasma in the neutral layer is often seen to drift tailward. The connection between auroral motions and the motion of magnetospheric plasma therefore indicate that the mag-

netic field lines at auroral latitudes are linked to the neutral layer somewhere deep in the magnetospheric tail, and that this region is closely related to the Solar Wind plasma.

The northern light is created by charged particles from the Solar Wind entering the magnetosphere and penetrating the Earth's atmosphere along magnetic field lines. These Solar Wind particles have energies much less than auroral particles but the acceleration processes which increase the energy of these particles are not fully understood. In connection with the creation of the northern light on the nightside, an explosive process takes place somewhere in the magnetospheric tail, and energetic electrons are blown out of the explosion center with considerable force. An electrical potential is often created between the magnetospheric tail and the ionosphere in the polar region accelerating the electrons to approximately 10,000 V. Electrons are forced to move along the magnetic field lines towards the polar region, and form two parallel illuminated ovals – the aurora borealis and the aurora australis. This process is shown schematically in Fig. 10.10.

10.8 The Auroral Oval

As already mentioned, Wargentin (Chap. 5.9) in the middle of the 18th century noticed that the northern light probably formed a ring around the pole and later Nordenskiöld in 1878–79 (Fig. 6.16) and Tromholt in 1885 (Fig. 6.10) demonstrated this ring. Nordenskiöld, however, noticed that the ring was not centered around the north pole, but rather around a point between the North Pole and the north magnetic pole, the so-called pole of the northern light. Lack of

Fig. 10.10. A simplified model of the formation of the polar light. An explosion-like process occurs in the tail of the magnetosphere, and electrons are forced to propagate along the magnetic field lines toward the polar regions. When these energetic particles hit the upper atmosphere they collide with atmospheric gases and form two parallel luminous ovals around the poles – the aurora borealis and the aurora australis

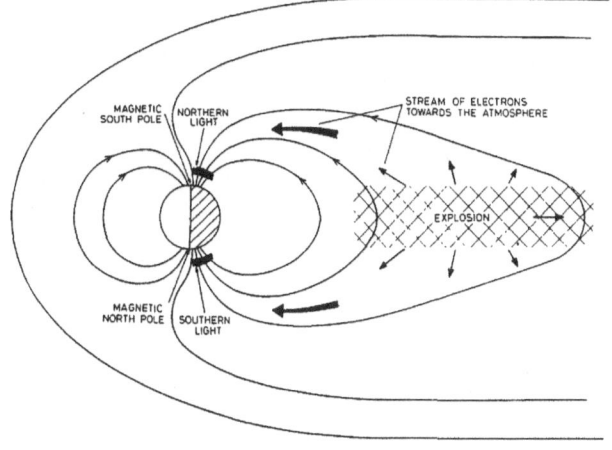

133

sufficient data made it impossible to characterize this ring in detail. Not until the International Geophysical Year (IGY) in 1957, when all-sky cameras were distributed in a large array in the polar area, could this task be completed. It was then in the beginning of the 1960's shown that the instantaneous northern light appears along an oval-shaped zone. In later years, however, it has been found that the polar cap area inside the oval actually is very close to being circular ring and centered at the geomagnetic pole.

This oval-shaped area is not a static feature but varies much in shape and position in accordance with the disturbances in the magnetosphere.

We are accustomed to thinking that the northern light only occurs at night, but there are also northern lights occurring on the dayside. These are caused by energetic particles penetrating the Earth's polar atmosphere directly from the Solar Wind in the so-called polar cusp or cleft regions (Fig. 10.7). These regions are characterized by magnetic field lines splitting into two parts – one branching across the polar cap toward the nightside and the magnetospheric tail, the other branching off towards the dayside and the front of the magnetosphere.

On the dayside Solar Wind particles penetrate the upper atmosphere in the auroral oval at a distance of approximately 10°–15° from the geomagnetic poles, while solar wind particles on the nightside are guided through the magnetospheric tail and into the high latitude atmosphere in the auroral oval at a distance of 20°–25° from the same poles.

These energetic particles collide with gas particles in the upper atmosphere when they approach the Earth along the magnetic field lines. Finally, when these collisions occur, polar light occurs as aurora australis and aurora borealis. Therefore these light phenomena are the finale of a long drama in which Solar Wind particles play the leading role.

10.9 The Northern Light and Auroral Particles

As already stated (Chap. 10.7) the Solar Wind is the primary source of auroral particles, as these particles create the northern light and cause radio disturbances and large variations in the Earth's magnetic field. Auroral particles penetrate down to about 100 km altitude and can therefore only be directly studied by the use of rockets and satellites. One very common experiment performed from satellites or rockets is to measure the energy of the incoming particles. If an electron passes through a potential drop of 1 kV the electron gains an energy of one kiloelectronvolt (1 keV). Such measurements of an electron's energy obtained simultaneously with measurements of the distribution of light emissions from the northern light show that 6 keV electrons are distributed in time and space in a similar manner as the auroral light (Fig. 10.11). The northern light on the nightside is usually created by electrons with energies of 5–15 keV while on the dayside it is caused by electrons with energies of 0.1–3 keV. Therefore, observations of the northern light can give information about where in the atmosphere such particle precipitation is important.

Auroral particles entering the upper atmosphere will sooner or later collide with molecules and atoms of the atmospheric gases. During each collision, electrons lose on the average 35 eV of energy. A 10 keV

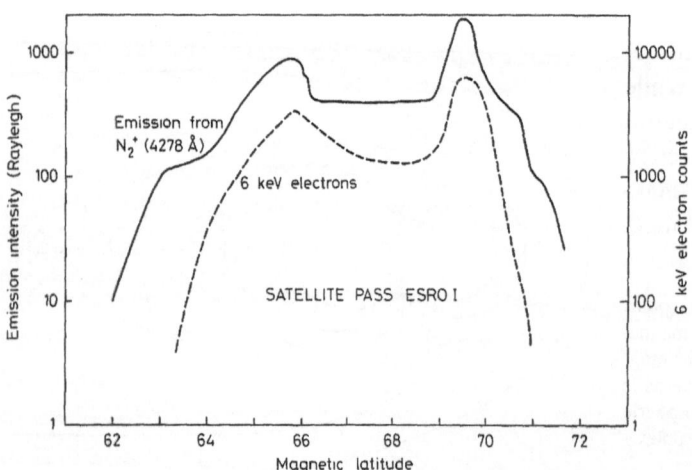

Fig. 10.11. The auroral emission at 4.278 Å and the energetic electron flux at 6 keV are similarly distributed in time and space, indicating a very strong relationship

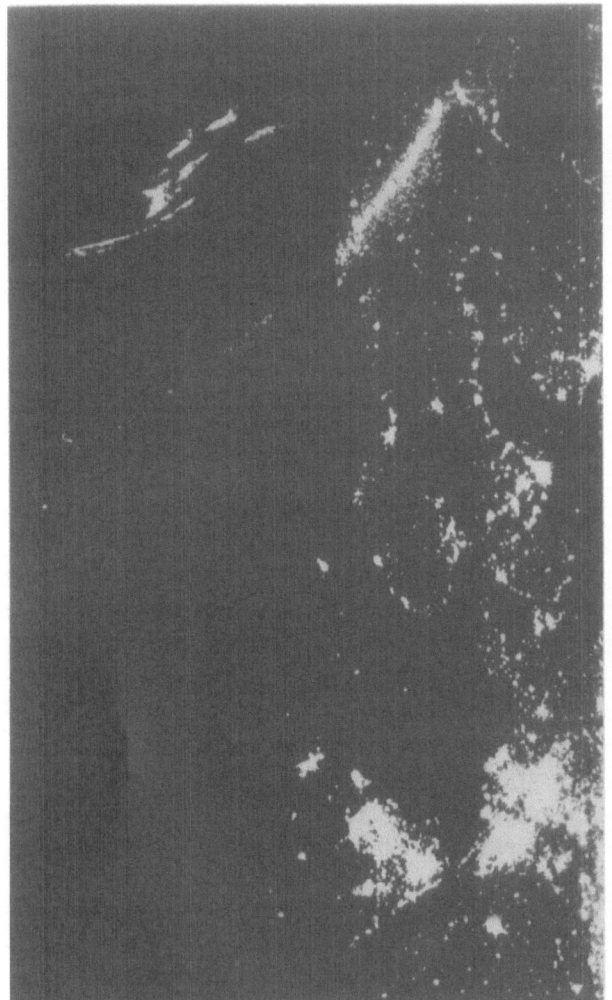

a

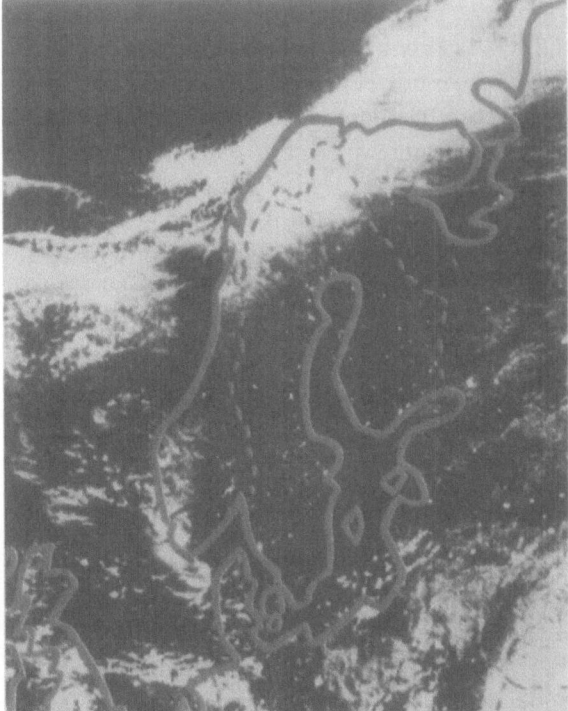

b

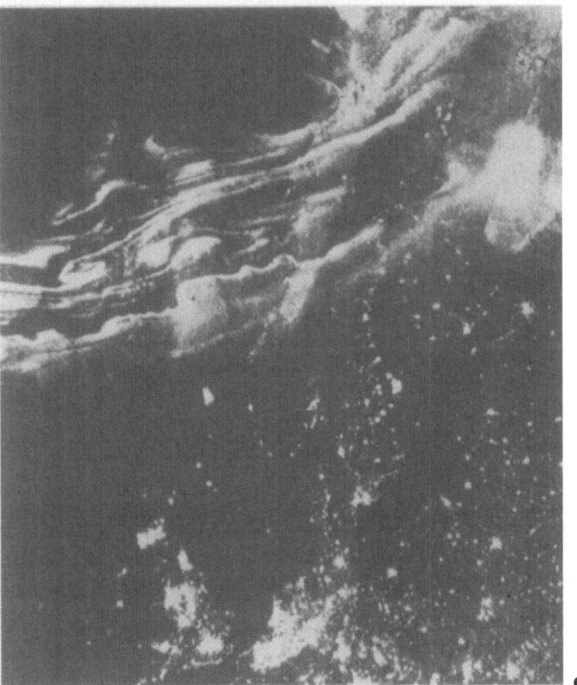

c

Fig. 10.12 a, b, c. The figures show the distribution of the northern light over Scandinavia for three different levels of activity. In a, part of the auroral oval is seen touching the northern tip of Norway. This is for a day of very low auroral activity. In b, the northern light has moved southward and almost reaching to the middle of Norway. This is a day of medium activity. The contours of the coastlines are drawn to help identify the position of the northern light. In c, the activity is very high and the northern light is reaching as far north as Trondheim (see Fig. 11.22) (DMSP photographs. Courtesy; Air Force Geophysical Laboratory)

electron can therefore undergo approximately 300 such collisions. An electron with an energy of 10 keV or a velocity of 60,000 km s^{-1} at the top of the atmosphere will be completely stopped due to collisions with atmospheric gases at 100 km above ground. Correspondingly, electrons with energies of 1 keV or a velocity of 18,000 km s^{-1} will be stopped above 130 km altitude, and a 0.5 keV electron will be stopped above 160 km. Protons must have more than 10 times as much energy in order to reach the same altitudes as electrons. The pitch angle between the particle's trajectory and the magnetic field is also important to the motion of the particles. A particle with small pitch angle or moving almost parallel to the magnetic field will penetrate deeper into the atmosphere than a similar particle with a large pitch angle. We therefore conclude that the height of the northern light is determined by the energy and the pitch angle of the penetrating particles (Fig. 10.12).

135

On the average only one photon is emitted at 4,278 Å (a very important auroral emission) for every 75 collisions between electrons and the atmospheric gas. In order to see the northern light with the naked eye more than 10^8 photons $s^{-1}cm^{-2}$ at 4,278 Å are needed, which means that more than 10^{10} collisions must occur per second in a column of air 1 cm^2 in cross-section in the upper atmosphere at altitudes above about 100 km. It is therefore clear that the intensity of the northern light is determined by the number of penetrating energetic particles.

By comparing variations in auroral intensity and the amount and energy of the penetrating particles it is possible to derive the ratio between light emissions and the amount of energy input to the upper atmosphere.

In order for the northern light to be visible to the eye, the energy input to the atmosphere must be about 1 $erg\,cm^{-2}\,s^{-1}$ or approximately $10^{-3}\,W\,m^{-2}$. For a medium strong northern light about 10 km wide and 1,000 km long, approximately 10^7 kW is needed which is comparable to the effect of a rather large hydro power plant. Since only 1% of the energy input to the atmosphere during an auroral display is used to produce visible light, it is clear that there is an enormous quantity of energy released in the upper strata of the atmosphere in the auroral oval during each auroral night.

10.10 Hannes Olof Gösta Alfvén Won the Nobel Prize in 1970 Partly for his Work on Auroral Physics

One major reason for neglecting the theories by Birkeland in the beginning of this century was the fact that if the Sun emitted electrons with negative charges only, it would soon be left so strongly positively charged that no further electrons could escape. The particle stream from the Sun would cease and northern lights would ultimately stop.

The Swedish physicist Hannes Olof Gösta Alfvén (1908–) postulated (around 1950) that the particle stream from the Sun is neutral in the sense that equal

Fig. 10.13. Hannes Olof Gösta Alfvén a Swedish professor and winner of the Nobel Prize in physics partly due to his theory of the northern light

negatively (electrons) and positively (protons) charged particles are emitted at the same time. Since these particles are moving in an interplanetary magnetic field, an electric field will be formed in the particle clouds streaming outwards from the Sun. When these clouds hit the Earth's magnetic field it is the electric field, which is carried along with them, which interacts with charged particles in the Earth's magnetosphere and, in fact, "open up" the magnetosphere to the solar particles.

Alfvén presented his theory in full in 1950 at a time when it was impossible to crucially test it. Not until the first satellites were in orbit around the Earth could Alfvén's ideas therefore be fully appreciated. The satellites showed clearly that an electric field is present in the Solar Wind and that the solar-wind particles are streaming through an interplanetary magnetic field.

Not only could Alfvén's theory very nicely explain how the solar wind particles can interact with the Earth's magnetic field, but also how an interplanetary electric field can penetrate into the Earth's magnetosphere and force the plasma around the Earth into motion in such a way that currents are formed and particles penetrate into the atmosphere to form the northern light.

Alfvén's theory is today the fundamental on which our understanding of the northern light is based.

11 The First Systematic Observations of the Northern Light in Norway – Auroral Observatories and Instrumentation

11.1 The First Auroral Observations in Alta, Northern Norway

In the 19th century a more systematic manner of studying the aurora in Norway came into being. As we look back on this fundamental pioneering work from our advantageous point of today, we can appreciate the ingenuity, the hardships, and the brilliance of these early auroral scientists and their important contribution to modern auroral research in the space age. In this chapter, we will briefly cover some of the first systematic auroral studies which were made, describe the important auroral expeditions into Northern Norway, and delve into the development and use of instruments for studying the main properties of the northern light.

In the years 1838–1839 a French expedition settled in Bossekop in Alta, in the most northern part of Norway. This was the French "Recherche" expedition under the directorship of Bravais and Lottin and was financed by the French king Louis Philippe who had visited this part of Norway as a prince. It was the intention of the expedition to observe auroral forms simultaneously at two different places in order to triangulate the height of the light emissions. The distance between the two observing points was too small, however, resulting in large uncertainties in the measurements.

Accompanying the scientists on this expedition was a very skilled artist, Bevalet, who together with Bravais made some of the most excellent paintings of the northern light which made the expedition more lastingly famous than anything else it accomplished. Four of their illustrations are shown in Fig. 3.3.

The director of the Norwegian Meteorological Institute, Henrik Mohn (1835–1916) played a central role in establishing the First International Polar Year (1882–1883). He initiated the installation of a field station at Bossekop in the care of Axel Severin Steen (1849–1915) who later became Mohn's successor as a director of the same institute. During the same year Sophus Tromholt (Chap. 6.5) installed himself at Kautokeino, 100 km to the south of Alta. The plan was to observe the same point of the aurora simultaneously from these two places and from these observations to estimate the altitude of the northern

Fig. 11.1. Bossekop, Alta, Norway. The first systematic auroral observations were carried out here in the last century. This is a picture made by Bevalet when he participated in the expedition to Alta in 1838–1839

Fig. 11.2. Sophus Tromholt, dressed in local Lappish costume, at the auroral station at Kautokeino, Finnmark. Tromholt participated in the First International Polar Year (1882–1883), and was responsible for the Kautokeino station. Notice also the different instruments he used

light. A special observation program was agreed upon beforehand, since the stations were widely separated and communication required days of time. From 18 simultaneous measurements Tromholt calculated the auroral height to be somewhat less than 113 km (71 miles). As this is a very reasonable value for the height of auroral arcs (see Chap. 7.3), that part of the programme was satisfactory. However, with this technique one can never be certain that both observers are really viewing the same auroral form. Furthermore, time synchronization at two places separated by more than 100 km was a problem in those days. Tromholt, however, was a man of many interests and he brought with him a photocamera to preserve for posterity some of the impressions of his grand tour. He also tried to photograph the northern light. This was probably the very first attempt ever made to photographically record this phenomenon.

In 1892 the German physicist Martin Brendel (1862–1939) visited Bossekop in an attempt to photograph the northern light. He succeeded in taking a few pictures of an auroral band by the use of a sensitive orthochromatic plate and an exposure time of only 7 s. These were probably the first photographs of the northern light published in a scientific journal.

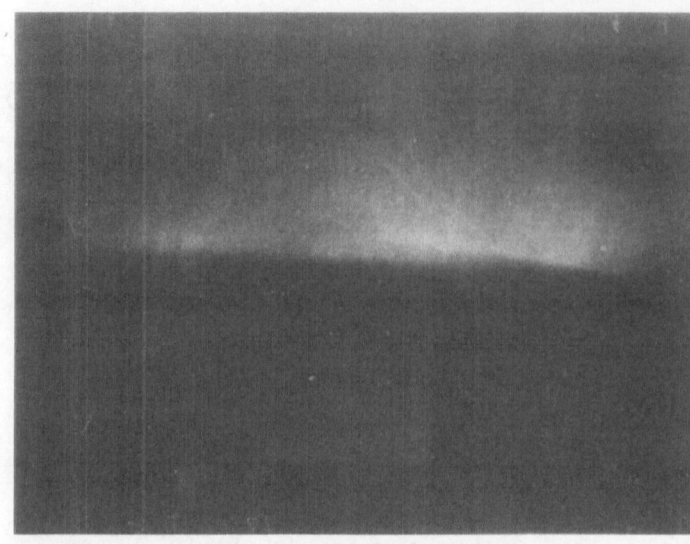

Fig. 11.3. The first successful published photograph of an aurora. This picture was taken in January 1892 by the German M. Brendel, at Bossekop, and the exposure time was only 7 s

Fig. 11.4. The first magnetic observatory at Haldde, built of stones in 1899 is shown in the background of this picture. This was the first permanent observatory erected in Norway. A similar building was put up at Talviktoppen, ~3.4 km from this station. They are both 900 m above sea level. The house in the foreground is the more modern Observatory which was built in 1912–1913. A tunnel was made between the two stations which made it possible to keep in contact even during stormy weather

Although the quality of Brendel's photographs was very poor compared to those of today, the introduction of photography opened possibilities of a far better way of measuring the height of northern light than by visual observations. The photographic method was to be fully exploited by Carl Størmer almost 20 years later (cf. Chap. 7.3).

11.2 Birkeland's Expeditions into North Norway

Brendel's success gave Birkeland the idea to triangulate the northern light by simultaneously photographing the same auroral form from two different places.

In 1896 Kristian Birkeland introduced an auroral theory (cf. Chap. 7.5) and he was eager to prove the correctness of it experimentally. In 1897 he travelled to Finnmark in search of a suitable observation site in the mountains between Bossekop and Katuokeino. A terrible snowstorm occurred which made it impossible to complete the expedition, but the next autumn Birkeland returned to Finnmark and checked out six of the highest mountains surrounding Kåfjord (Alta). He chose two of these, Haldde and Talviktoppen both about 900 m high, as being suitable. In 1898 he applied for money from the Norwegian government to build an observatory on the top of each of these mountains. In the summer of 1899 they were both finished, and Birkeland spent part of the winter 1899–1900 on Halddetoppen (see Fig. 11.4).

One of Birkeland's main goals in the first expedition was to measure the height of the northern light using the photographic technique. Also, a telephone cable between the two mountains made it possible to expose photographic plates simultaneously at both stations.

In addition to taking photographs for use in measuring auroral heights, Birkeland also recorded variations in the Earth's magnetic field. Preliminary studies of these geomagnetic observations convinced Birkeland of a very close relationship between auroral displays and geomagnetic disturbances at Haldde and the simultaneous geomagnetic disturbances at Potsdam. On the basis of his observations he wrote the book *Expédition Norvégienne de 1899–1900 pour l'étude des aurores boréales. Résultat des recherches magnétiques* dedicated to his assistant Elisar Boye

who perished in a snow avalanche on his way from Kåfjord to Haldde.

In order to find the real nature of the electric currents related to the auroral displays and the geomagnetic disturbances Birkeland was convinced that he had to expand his network of observatories. Therefore he equipped a new expedition during the winter of 1902–1903 and placed his assistants at four field stations in the Arctic. The stations were at Kåfjord (Haldde) in Finnmark, Dyrafjord in Iceland, the Axel Iceland at Svalbard (Spitzbergen) and at Novaja Semlja. To work on the enormous data material from this expedition Birkeland engaged Lars Vegard (Chap. 7.4) and Ole Andreas Krogness (1866–1934) who later became the director of Haldde. The first results of this expedition were published in the gigantic work *The Norwegian Auroral Polar Expedition 1902–1903*. Volume I, part one was printed in 1908 and part II in 1913.

The technique of such a network of stations, as that introduced by Birkeland 80 years ago, is widely used in auroral and geomagnetic studies today.

11.3 Norway's First Permanent Auroral Observatory on Haldde in Kåfjord

In 1910 Halley's comet, which had not been seen since 1834, once again appeared. Birkeland was interested in this event because he believed that the tail of the comet carried charged particles from the Sun which could lead to magnetic and/or electrical disturbances on the surface of the Earth. In May of 1910 he went to Haldde to study the effects of Halley's comet because he believed that these effects would be most significant in the far north. The results from this expedition were so promising that Birkeland was convinced that a permanent observatory should be built on Haldde.

Birkeland received 30,000 Norwegian kroner (Nkr) for the buildings and 10,000 Nkr to operate a magnetic-meteorological observatory on Haldde. He believed that observations covering two sunspot periods, i.e., 22 years, were necessary in order to obtain meaningful data and thus the money should be provided for this same time period. In his application to the Norwegian government Birkeland emphasized the practical usefulness that a permanent observatory on Haldde would have for weather forecasting in Northern Norway. It should be pointed out that the fishing industry is very important in this part of the country and a reliable weather forecast would be very valuable for the fishing export industry. This is probably the reason that the government agreed to build a new observatory, and it also illustrates Birkeland's political acuity as being equal to that of present-day auroral scientists. The observatory (cf. Fig. 11.4) was built in 1912–1913 and early in the autumn of 1912 Ole Krogness became its first director.

11.4 The First Plans
for a Geophysical Institute in Tromsø

In the autumn of 1915 Ole and Dagny Krogness were joined by another enthusiastic couple, Olav and Dagny Devik. In 1911, Olav Martin Devik (1886–) became one of Birkeland's assistants working on the famous Terella experiment (cf. Chap. 7.5) and, beginning in 1914, he was engaged in plans for founding a permanent weather forecasting system in Northern Norway. Because of the First World War from 1914–1918 (although Norway was neutral) it was inconceivable to plan another permanent observatory in Northern Norway in addition to the one existing in Haldde.

The first plan for a geophysical institute to be established in Tromsø was nevertheless presented by Devik and Krogness in 1915. Based on their experiences at the Haldde observatory, they believed it would be almost impossible to keep bright young researchers at such an isolated place. It had also become evident that it was not mandatory to make the observations from such great altitudes as the Haldde. Observations of equal quality could be made at Tromsø, an old Norwegian commercial seaport.

Krogness and Devik began an intensive campaign, in which they were very successful, to raise money and support for a new geophysical center in Tromsø. In addition, the town of Tromsø donated a suitable plot of land for the location of the buildings.

The Society for a Geophysical Institute in Tromsø was established early in 1917, and in May of the same year the Norwegian Parliament unanimously approved the issue with the following words: "The National Congress gives its approval to the establishment of a Geophysical Institute in Tromsø" (cf. Fig. 11.5). This new institute was also to be responsible for the operation of the Haldde Observatory which was a solution that Krogness and Devik especially looked forward to. When they finally moved into the new building in December 1918, they could look back to 6 and 3 years, respectively, spent at Norway's most weather-beaten site – the Haldde.

The Swedish physicist Hilding Köhler (1888–) was appointed as the new manager of the Haldde Observatory – a position which he held until the summer of 1926. The permanent observations at Haldde ceased in 1927 because operating expenses had become too great.

Krogness was appointed as first director of the new institute at Tromsø while Devik became the leader in the new field of research – weather forecasting.

11.5 The Auroral Observatory in Tromsø

Naturally, the work related to weather forecasting was heavily emphasized at the new Geophysical Institute in Tromsø and as a consequence basic research, with emphasis on northern light and geomagnetic phenomena, was given a lower priority.

Another of Birkeland's assistants, Lars Vegard (see Chap. 7.4) was primarily interested in spectral studies of the northern light. Vegard had obtained instruments of greater dispersion and higher quality than had been available previously (see Chap. 11.6e) and he mounted this equipment on the roof of the

Fig. 11.5. The Geophysical Institute in Tromsø – shortened to Geofysen by the local people – was completed in 1919. This soon became the centre for weather predictions in Northern Norway, but studies were also done on the northern light and the Earth's magnetic field. A platform was mounted on the roof from which the auroral spectral measurements were made

Geophysical Institute. However, the space available on the roof for Vegard's research quickly became too small and he began looking for other space.

On February 24, 1925, he applied for funding from The International Education Board (or the Rockefeller Foundation) for an auroral observatory to be located in the vicinity of Tromsø.

In May 1927, the Rockefeller Foundation approved the plans for an auroral observatory to be established in Tromsø and granted $75,000 to be used for building and equipment. The condition of the grant was that the Auroral Observatory, with equipment and land, was to be owned by Norway. Furthermore, the observatory was to be run in compliance with rules and laws approved by the National Congress, and the Government should provide operating funds of approximately 12,000 Nkr annually.

The observatory was built on a 10-acre tract of land donated by the town in the most suitable area of Tromsø Island (cf. Fig. 11.6). The building was finished in 1928 providing Norway with a modern auroral observatory. The first director was Leiv Marius Harang (1902–1970) who started his work there in the summer of 1928 although the official opening was not held until 1930.

Fig. 11.6. A picture of the Auroral Observatory in Tromsø, taken early in the 1930's. It was mainly thanks to a grant of $75,000 from The Rockefeller Foundation which made it possible to establish a modern observatory in the end of the 1920's. The original buildings are still in use, but several new ones have been erected. They are all now integrated into the University of Tromsø

During the first 10 years at the Auroral Observatory much research was accomplished by the young research staff and their dynamic leader, Leiv Harang (Fig. 11.7). One has only to glance through the literature for that period to see the important scientific papers which were published by this group. In particular during The International Polar Year of 1932–1933 many famous scientists spent various periods of time doing special studies at the Observatory. It was during this time that the German engineer Willy Stoffregen (Fig. 11.8) first became associated with the staff (see Chap. 11.6 b). Stoffregen proved to be a very inventive worker and together with Harang did pioneering work in the field of radio propagation studies. He also built instruments which were to become standard in auroral research, some still being in use today. Stoffregen was forced to flee from Norway when the Second World War broke out and

141

Fig. 11.7. Professor Leiv Marius Harang (1902–1970). Harang became the first director in 1928 and worked at the Auroral Observatory for almost 18 years. In 1946 he became superintendent for the Division of Telecommunications at the Norwegian Defence Research Est., and in 1952 he was appointed professor at the University of Oslo. Harang remained through his life an active supporter of the research program at the Auroral Observatory in Tromsø

Norway's interests and accomplishments in the field of geophysics during the 1950's and 1960's were instigated by Harang. He understood the problems of combining available resources and bridging gaps between different fields of research which resulted in international recognition for Norwegians.

In the 1960's, plans for a university in Tromsø gained momentum, and in 1970 it was finally decided to build the most northerly located university in the world. Simultaneously, it was decided that the Auroral Observatory should be fully integrated into the university as part of the Institute of Mathematical and Physical Sciences.

Norway was occupied by the Germans. During these 5 war years of German occupation very little scientific work was done by the staff. However, the staff actively participated in illegal work involving radio communication with the Norwegian government in England.

As time went by, the Germans became suspicious of this activity at the observatory and in January 1945 Harang was arrested and taken to Germany where he was imprisoned until the end of the war.

Leiv Harang worked at the Auroral Observatory for almost 18 years, and mainly concentrated his efforts on the physical processes occurring in that part of the upper atmosphere known as the ionosphere. In his studies he was concerned with both radiowaves, auroral observations and earth magnetic field measurements. Soon Harang's name became internationally well known in auroral and geomagnetic research. Even though in 1946 he moved from the observatory and Tromsø to Oslo, he remained an active supporter of the research programme at the observatory for the rest of his life.

11.6 Basic Instruments for Auroral Studies

a) The Krogness-Størmer Camera. All auroral observations before the beginning of our century were

Fig. 11.8. Dr. Willy Stoffregen (1910–) did pioneering work in the field of radio propagation studies together with Leiv Harrang. Stoffregen is well known internationally for his inventive work in geophysics

Fig. 11.9. This is the Krogness-Størmer camera which they together developed for auroral use in 1908. Internationally, it is usually referred to as Størmer's auroral camera, but Krogness played a major role in its development. The camera on the right – viewed from the back – is a more advanced unit (from roughly 1915) by which it was possible to take six pictures on the same plate by moving the lens

severely restricted because of the use of out-dated instruments and the strong attraction to personnel of old traditional views. Since 1909 the photographic camera has been used extensively as a primary auroral measuring instrument.

The Norwegian auroral camera was very popular with research people all over the world until the Second World War. The number of auroral photographs made with this camera is enormous, and in Norway alone the number of such auroral pictures exceeds 100,000. A picture of this camera is shown in Fig. 11.9, and some auroral forms photographed with it are shown in Fig. 3.1.

With such simple cameras the first accurate systematic studies of the northern light were begun (cf. Chap. 7.3).

b) Automatic Auroral Cameras with 180° Field of View. In the middle of the 1950's a new and very useful camera was designed for auroral research. With the camera shown in Fig. 11.10 it is possible to take an all-sky picture of the northern light every minute during darkness, under favourable weather conditions. The required time of exposure is determined by the type of film but usually it is 10–30 s for the fastest films. This means that rapid movements in the northern light are smoothed out during this relatively long exposure time but it does permit one to study auroral activity throughout the night, minute by minute, from the horizon to the zenith.

This camera automatically photographs the sky and accurately records the time and orientation. A few auroral pictures taken with the all-sky camera are shown in Fig. 11.11. The upper mirror (see Fig. 11.10) obscures a little part of the sky in the zenith (the dark, circular area in Fig. 11.11) but this is corrected for in modern all-sky cameras by the use of fish-eye lenses rather than mirrors.

More than 100 all-sky cameras are in continuous use today in the auroral zones. The main aim of this study will all-sky cameras is to map the frequency of auroral occurrence as a function of place, time, and

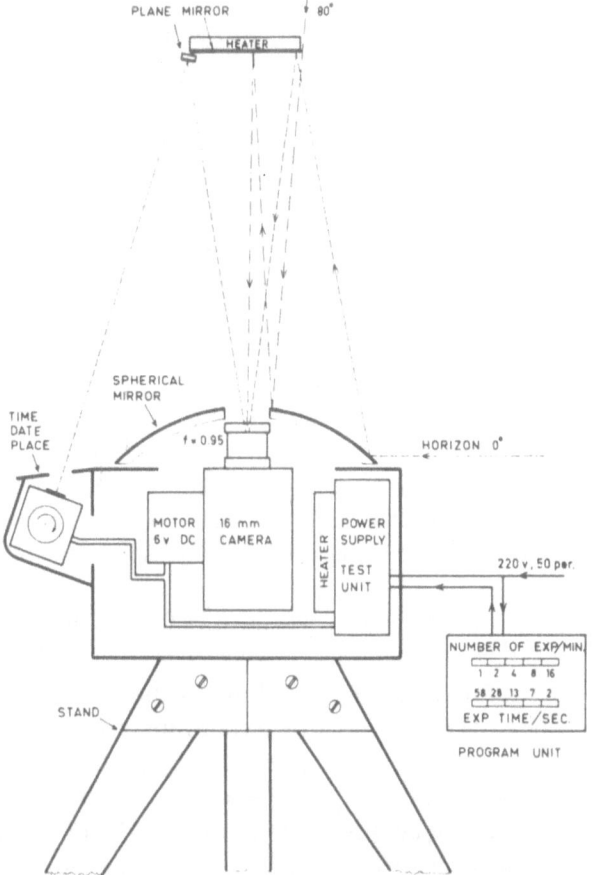

Fig. 11.10. A block diagram of Stoffregen's all-sky camera. A convex spherical mirror reflects an image of the sky on to the plane mirror at the top. The 16-mm movie camera, in the main housing, photographs this picture through the opening in the mirror's center. With this arrangement, the camera has an 180° field of view and also both direction and time are included. In the earlier all-sky cameras, 16 mm film was standard but today 35 mm or larger films are also used

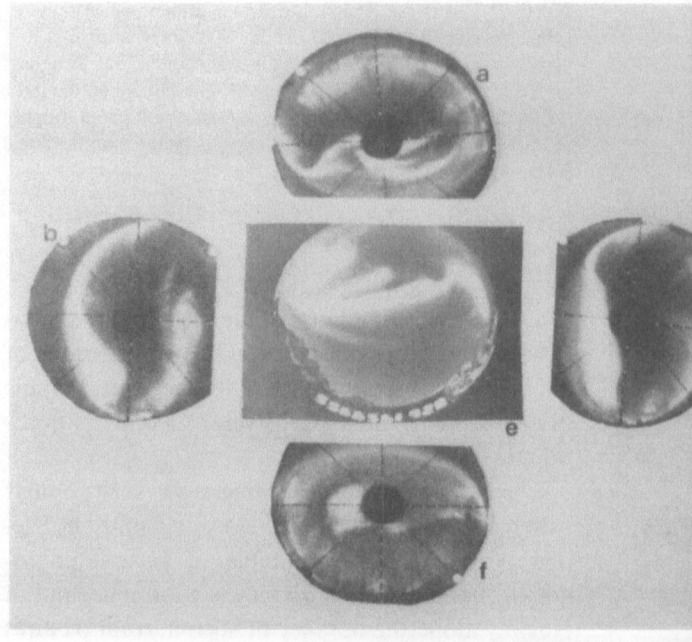

Fig. 11.11. Some pictures taken with modern (centre) and older type all-sky cameras. Today wide-angle, fish-eye lenses are used to image the upper hemisphere of the sky. The exposure times are normally between 10 and 30 s

solar activity. It is also possible to study the occurrence of different types of auroral forms and to obtain information about the movement of the northern light.

c) Amateur Auroral Photography. On a clear, dark winter night the northern light may look very bright to the dark-adapted naked eye, and this could be deceptive for it is not easy to take good pictures of a northern light. With a fairly modern 35 mm camera with wide-angle, small focal ratio lens, however, it is well possible to try. With a fast colour film (e.g., 400 ASA) an exposure time of a few seconds is normally needed and therefore it is well advised to use a tripod for mounting the camera.

d) Television Auroral Cameras. In the middle of the 1960's closed circuit TV-cameras were first introduced into the study of the northern light. The expanding development in modern electronics during the last decade has resulted in more and more TV-systems being used in auroral research. The great advantages of the TV-camera are the short exposure time and its much greater sensitivity than photographic film. With an auroral TV-camera 50 pictures per second are taken which makes it possible to carefully study rapid variations in auroral forms. With the sensitivity of this equipment it is possible to study in detail a northern light which is invisible to the naked eye.

Every TV-picture is composed of approximately 600 thin horizontal lines of which two adjacent lines are sufficient to give useful information about the object. Therefore, such TV pictures are very useful in studying fine structures (small-scale structures) in the northern light. With this equipment it is possible to study auroral structures down to the order of 10 m.

Using lenses of different focal lengths different fields of view can be obtained, which make it possible to concentrate on widely different auroral forms. TV-auroral cameras are particularly well suited to the study of auroral pulsations, which vary so rapidly. TV recordings are made on video tape which are easy to copy and distribute to other laboratories for correlative studies. A disadvantage of the TV-cam-

era is that light intensity determined with this technique is much less certain than with a photomultiplier tube.

e) Instruments for Studies of Colours in the Northern Light. In 1915, L. Vegard (Chap. 7.4) built some advanced, very sensitive prism spectrographs to determine the different colours of the northern light. The principle of the spectrograph is that auroral light is separated according to colour by the use of a prism as illustrated in Fig. 11.12. The colours spatially recorded on film make it possible to measure accurately the wavelengths of different auroral emissions. This instrument with its associated lenses, prism (or grating) and film is called a spectrograph.

The first spectrographs were simple in design and their sensitivity was very low. In order to obtain a spectrogram on the photographic plates, one often needed an exposure time of at least 1 h. The amount of light collected by a prism is determined by the opening, or the area, of the collimator lens and the slit width. To gather more light the opening had to be increased, but this in turn reduced the wavelength resolution (i.e., the separation of two spectral lines lying close together). It was therefore necessary to make a compromise between exposure time and wavelength resolution.

Today's auroral spectra recording equipment uses modern electronic and data techniques, instead of photographic material, for storing information.

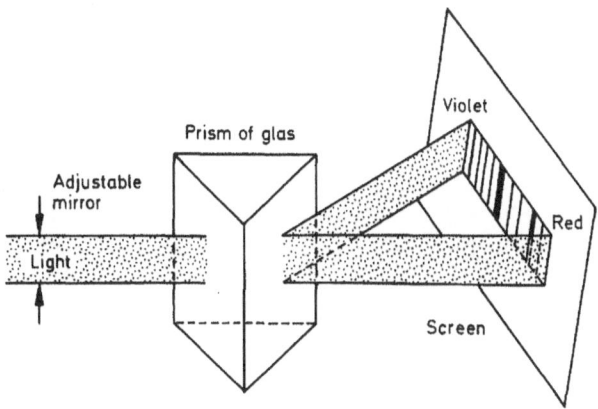

Prism of glas

Violet

Adjustable
mirror

Red

Light

Screen

Fig. 11.12. This sketch shows a simple, prism spectrograph of the type used until the 1940's. The light which passes through the prism is refracted into the different colours and recorded on a photographic plate

Prisms have been replaced by extremely fine ruled gratings; i.e., mirror plates with a large number of parallel lines (up to 12,000 cm^{-1}). Such gratings give very good wavelength resolution. Often a minicomputer is directly coupled to the spectrograph, and thus the data are analyzed briefly after they are recorded. Presently, more information concerning an auroral spectrum may be obtained in one second than Vegard – 60 years ago – was able to gather during a whole winter season. Some examples of Vegard's data from 1940 are shown in Fig. 7.10, while an auroral spectrogram from a modern instrument is shown in colour on p. 122.

f) Auroral Photometers. Presently, an auroral photometer is the most convenient and accurate instrument for intensity measurements (the absolute brightness of the auroral form) as a function of wavelength. A photometer measures light intensity at a wavelength of interest by using a narrow band interference filter. The photometer's field of view is determined by a combination of lenses and shutter, and usually the opening angle is between 1° and 10° (cf. Fig. 11.13). The basic element of this instrument is the photomultiplier tube. When auroral light (photons) strikes the photocathode of this tube, electrons are released and are then accelerated by an electrical potential through the tube. The number of electrons freed by the photocathode is proportional to the light intensity which strikes it. This stream of electrons (the current) is amplified and recorded, either on a pen recorder or on tape. A photometer is, in principle, simply a very sensitive lightmeter.

If simultaneously one wishes to study several different colours occurring in a northern light then any number of photometers may be mounted together in one instrument. Auroral location and movements may be recorded by using a rotating mirror which reflects light from horizon to horizon into the photometer.

Thus, photometers are readily seen to be convenient instruments for studies of auroral movements over the whole sky. Photometers are also ideal for studying auroral pulsations since they are able to record changes in light intensity which occur in the order of a one-millionth of a second. However, with the extremely narrow field of view usually employed, the aiming of the instrument becomes very important if only a small region of the northern light is pulsating.

11.7 Ground Based Measurements of the Earth's Magnetic Field and the Upper Atmosphere

a) Magnetometers. In order to accurately record variations in the Earth's magnetic field one uses magnetometers. The basic part of this instrument is a permanent magnet which is mounted in such a manner that it can rotate freely in phase with changes in the Earth's magnetic field. Three components of the field are necessarily measured to determine the geomagnetic field. A sketch of a magnetometer is shown in Fig. 11.14. A small mirror is mounted, together with

Fig. 11.13. Sketch of a simple auroral photometer, used both on the ground and in space platforms. *Numbers* indicate the different elements. *1* electrical connector; *2* on/off switch electrical; *3* amplifier; *4* high voltage supply; *5* calibration unit; *6* high-voltage divider; *7* photomultiplier; *8* thermistor; *9* rocket/satellite structure; *10* shutter; *11* calibration lamp; *12* lens; *13* interference filter; *14* protection glass

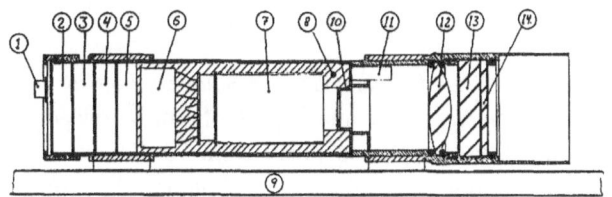

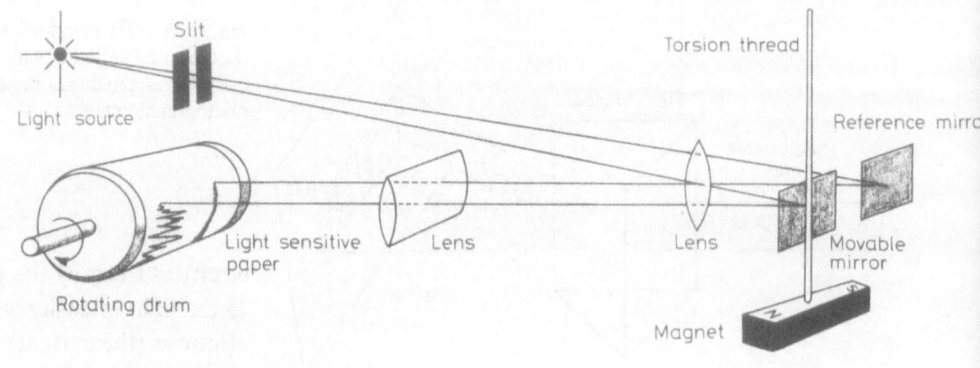

Fig. 11.14. a, A sketch of a magnetometer. The permanent magnet rotates when the Earth's magnetic field changes. A mirror, which rotates with the magnet, reflects and focuses the light beam on the continuously moving photographic paper. A fixed mirror gives a zero reference line. Using this technique, recordings of the type shown below are obtained. *b,* These curves illustrate a disturbance (a storm) in the geomagnetic field at Ny-Ålesund, Bjørnøya and Tromsø which started roughly at 1930 h

the magnet, on a torsion balance as shown. The mirror reflects a light beam and focuses it on to continuously rotating photographic paper. In addition, a fixed mirror reflects the same light to a steady position, giving a base of zero reference line.

Changes in the geomagnetic field result in rotation of the magnet along with the mirror. This is recorded on the photographic paper as a deviation from the baseline. A typical magnetogram is shown in Fig. 11.14 b. It is those deviations from the baseline, of course, which are of interest in investigations of the Earth's magnetic field, and which are of particular interest in connection with auroral observations.

Observation of the Earth's magnetic field is carried out at more than 200 observatories all around the world.

During the last 20 years, new magnetic instruments have been developed. They are based on different physical principles than the old type as shown in Fig. 11.14 a. Some are based on the fact that changes in the Earth's magnetic field strength will cause changes in the energy levels in rubidium atoms, some on measurements of magnetic flux through a coil and others on measurements of variations in the magnetic moment of protons. Instead of photographic recordings, modern electronics and on-line computers are often used. The greatest advantage of the latest recording devices is that curves, such as those shown in Fig. 11.14 b, are made available in real time. With the old technique one was forced to wait up to 24 hours until the paper was developed.

b) The Use of Radiowave Techniques; Ionosondes and Riometers. The main part of our knowledge about the upper atmosphere from 70–400 km has been obtained by radiowave techniques. The most

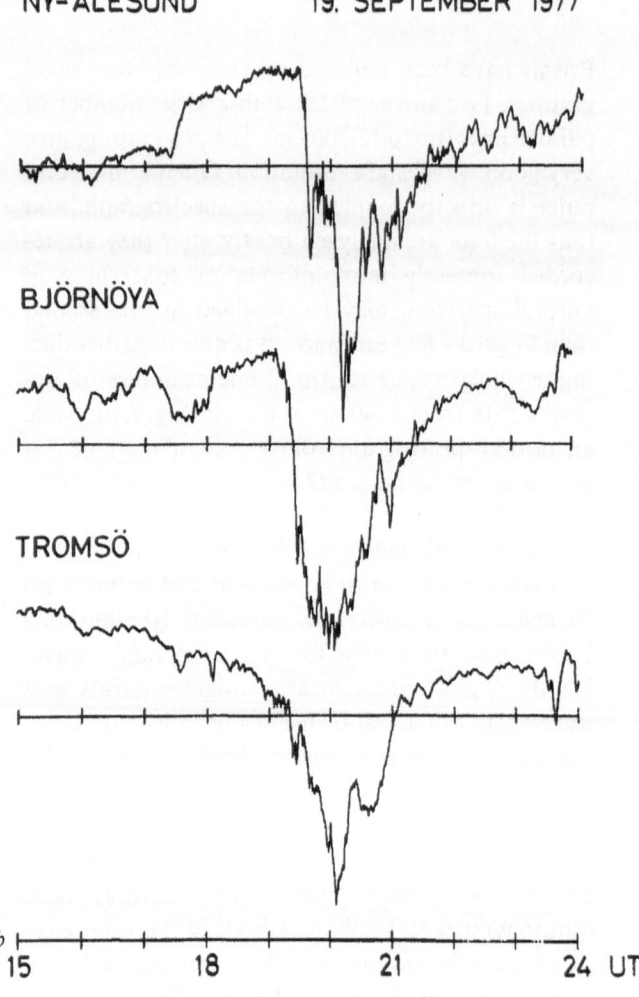

important instrument for studies of this height region has been, and still is, the ionosonde – the radio echo method. The ionosonde is a radar which transmits radio signals of different wavelengths (i.e., from 15–300 m) vertically upwards. A radio receiver records the reflected echoes from the ionosphere and measures the delay time. The sketch in Fig. 11.15 shows the main elements of an ionosonde. Instead of

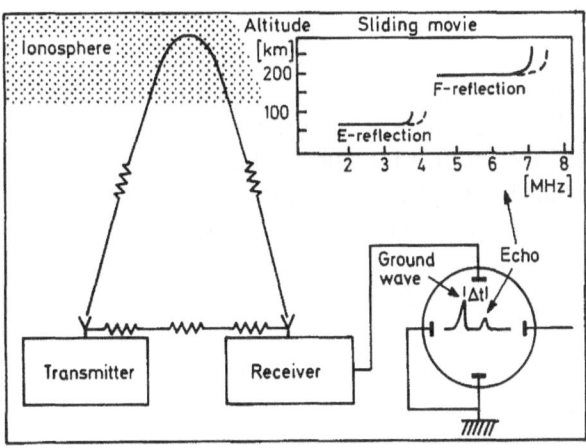

Fig. 11.15. A sketch of an ionosonde, an important instrument used in ionospheric research. The apparatus – a radar device – measures the height of the ionosphere. Radiowaves of different frequencies (1–20 MHz) are transmitted vertically, and the reflected waves are recorded photographically on moving film. The result, shown in the upper right corner, is an ionogram. By measuring the time difference between the direct and the returned wave, we get information about the height. The frequency contains information concerning the number of free electrons in the ionosphere

recording photographically, an oscilloscope display is used. With this display, the time difference between the direct and the returned echoes, Δt, is measured accurately as a function of wavelength. In practice, a film moves transversely across this screen and records the time delay of the echoes (relative to the ground wave). The ionogram shown in the upper right corner was obtained in this way. With this technique electrical properties of the upper atmosphere are continuously recorded from more than 200 stations all around the world. From Δt we get information about the height of the reflecting layer while the wavelength contains information about the number of free electrons existing at the different heights.

A number of other different radio wave experiments are used to obtain information on the electric properties of the upper atmosphere. One other fairly old technique is to measure absorption, i.e., the reduction in amplitude after the wave has propagated a known distance in the ionosphere. Absorp-

tion (mainly due to collisions between free electrons and neutral molecules) is primarily limited to the lower part of the ionosphere; i.e., between 70 and 110 km. Strong absorption events normally occur during periods of intense auroral displays and are therefore of considerable interest because they contain information about physical processes happening during these events. However, during such events all radio waves are completely absorbed, i.e., we have no reflection or return of waves from the ionosphere. Fortunately, one very convenient source is still available, radio waves from outer space, i.e., the cosmos. The origin of these cosmic radio signals is still not understood, but they have the character of random noise. These waves can easily be recorded on the ground on frequencies above the critical frequency for the ionosphere, $f > 20$ MHz.

The received power at a given frequency is approximately constant. With a reasonable directive antenna and a sensitive stable radio receiver this cosmic radio noise can be recorded. The received power, when no northern light is present, forms a baseline, i.e., the ionospheric absorption is practically zero (a quiet day curve). During strong auroral events, very little of this cosmic noise can penetrate the ionosphere. By calculating the difference between this and the quiet day level, the total amount of absorption due to the event can be estimated. From this value it is possible to get more information about the electric properties of the ionosphere below approximately 100 km.

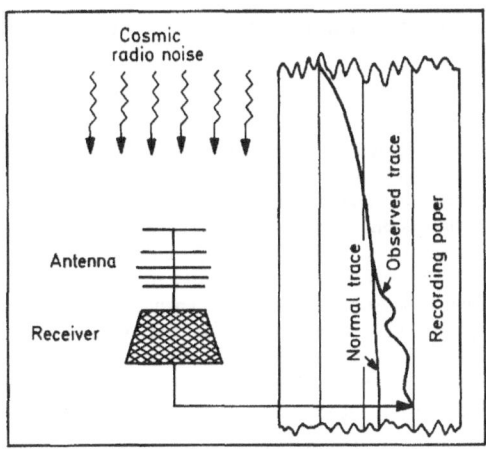

Fig. 11.16. This sketch illustrates the basic principle of an instrument for measuring radio wave absorption in the upper atmosphere. By recording variations in the cosmic radio noise, it is possible to obtain information about important electrical characteristics of the upper atmosphere

147

11.8 The European Incoherent Scatter Facility (EISCAT)

The European Incoherent Scatter Facility in Northern Scandinavia (EISCAT) is an inter-European research association where Finland, France, Norway, Sweden, United Kingdom and W. Germany participate. The facility consists of two gigantic radars placed at Ramfjordmoen, close to Tromsø in Northern Norway. Receiver stations are also installed at Kiruna, Sweden, and Sodankylä, Finland.

In order to comprehend the scientific interest in this facility it is necessary to present a brief outline of the physical ideas behind the use of high power and high frequency radars in ionospheric research.

Radiowaves with frequencies between 1 and 30 MHz (10–300 m wavelengths) when propagating vertically into the ionosphere from the ground, will reach an altitude where the wave frequency is equal to a characteristic frequency of the ionospheric medium, called the plasma frequency. At this point the radiowave will be totally reflected back to the ground. This plasma frequency increases by height in the altitude region between 100 km and 400–500 km,

and therefore radiowaves with the higher frequencies will be reflected from greater heights. The reflected signal, when decoded, can yield information about the electron density and the wind speed in this region of the upper atmosphere. This technique, which is called ionospheric sounding, has been widely used and is the most common method in ionospheric research (Chap. 11.7).

Above the height of maximum plasma frequency, however, it is impossible to get radiowaves, propagating from the ground, to be totally reflected. In order to study the top of the ionosphere with radiowaves in the frequency range of 1–30 MHz, radio transmitters and receivers have been installed in satellites, so-called topside sounders.

Since most of the power is preserved in a radiowave suffering total reflection, the ionosphere acts almost like a perfect mirror for such waves; it is therefore often called a coherent reflection or scattering. This is the reason why radiowaves in the frequency region 1–30 MHz can be used for long distance communication, by multiple reflections from the ionosphere and the ground (cf. Figs. 9.5 and 9.6).

For radiowaves with frequencies higher than 30 MHz, most of the power will penetrate through the ionosphere and into space, and no total reflection occurs. Any radiowave, propagating through the ionosphere, will lose power by interacting with free electrons in the wave path such that the electrons will gain a small fraction of energy by being forced to move with the oscillating electric field in the wave. An electron accelerated by such an oscillating field will, however, reradiate some of the power gained from the wave. This reradiated energy, however, will be scattered in all directions and some of it will also reach the ground. The electrons in the ionosphere are free to move more or less independently, such that the scattered energy coming back to the ground will, unlike the coherent total reflection, be an incoherent reflection.

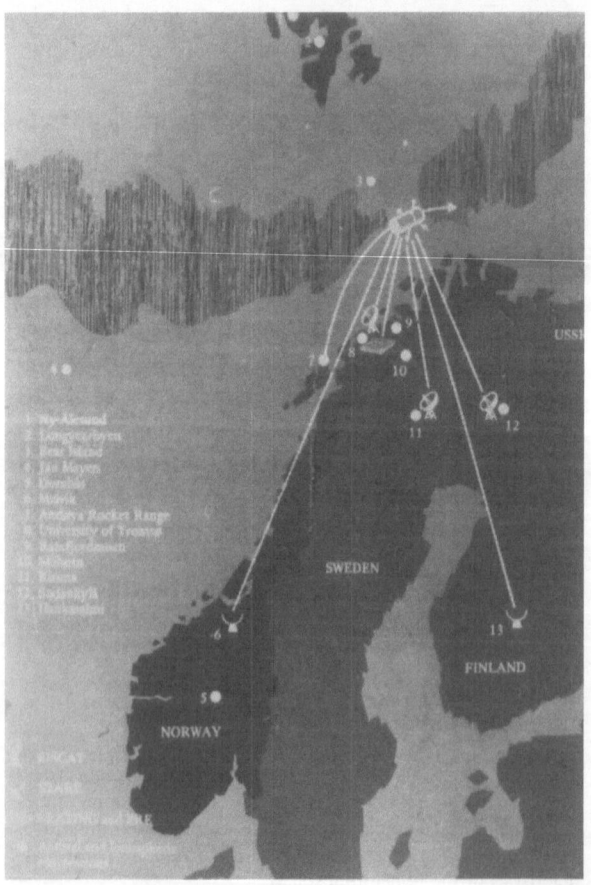

Fig. 11.17. An illustration showing the position of some of the installations for auroral research in Scandinavia. 1 Ny-Ålesund, permanent polar field station; 2 Longyearbyen, temporary polar field station; 3 Bear Island (Bjørnøya) permanent polar field station; 4 Jan Mayen, permanent polar field station; 5 Dombås, permanent field station; 6 and 13. The STARE (Scandinavian Twin Auroral Radar Experiment); 7 The Andøya rocket range and field station; 8 The Auroral Observatory at Tromsø; 9 The Heating system at Ramfjordmoen; 10 Skibotn auroral field station; 11 Kiruna Geophysical Institute; 12 Sodankylä Auroral Observatory; 9, 11 and 12 EISCAT, European Incoherent Scatter Facility

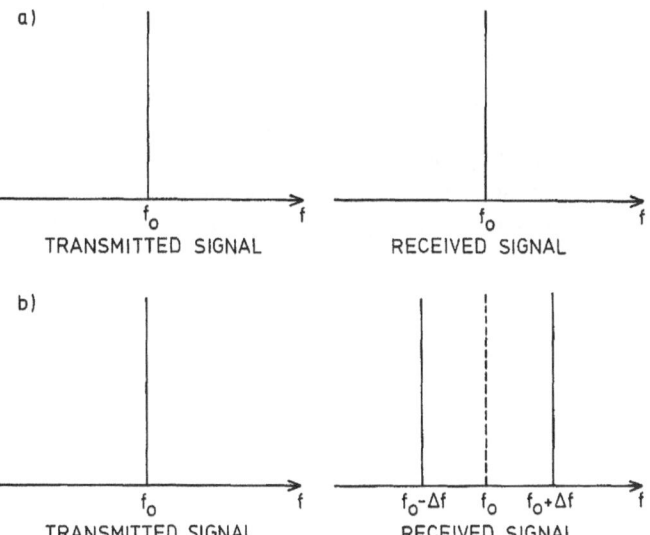

The difference between a coherent and an incoherent signal can be illustrated by comparison with a choir with and without a conductor, respectively. For a choir with a conductor the song will be strong, unisonous and clear, while the song from a choir without a conductor will be weak, random and incomprehensible.

An incoherently scattered radio wave is then very weak, and large installations are needed to observe this effect at all. Firstly antennas with very large collecting areas are needed, but since the collecting area of an antenna used for one radio frequency is four times as large in physical size as an antenna with the same collecting area used for twice the frequency, reasonably sized antennas can be obtained only for very high frequencies, 50–1,500 MHz. In addition to large antennas, strong radio transmitters of the order of 1–5 MW are also needed, in order to obtain an incoherently scattered signal which is strong enough to be observable from the ground. This explains the physical dimensions of the incoherent scatter facilities in use (see Fig. 11.20).

When a radiowave of one frequency is reflected from a target at rest with respect to the transmitter and the receiver, the wave will be received on the same frequency as it is transmitted; if, however, the

target is moving away from or towards the receiver, the received wave will have a frequency which is either lower or higher than the transmitted wave. The frequency difference between the transmitted and the received signal varies linearly with the target velocity along the line of sight as seen from the receiver (Doppler displacement) (Fig. 11.18).

In the ionosphere the electrons are moving independently in all possible directions and with all possible speeds, therefore the scattered signal will be spread out in a broad frequency range (spectrum) around the transmitted frequency. The shape of this spectrum carries a large amount of information about the physical conditions of the ionosphere. The electrons are negatively charged and cannot exist freely in the ionosphere unless there is an equal

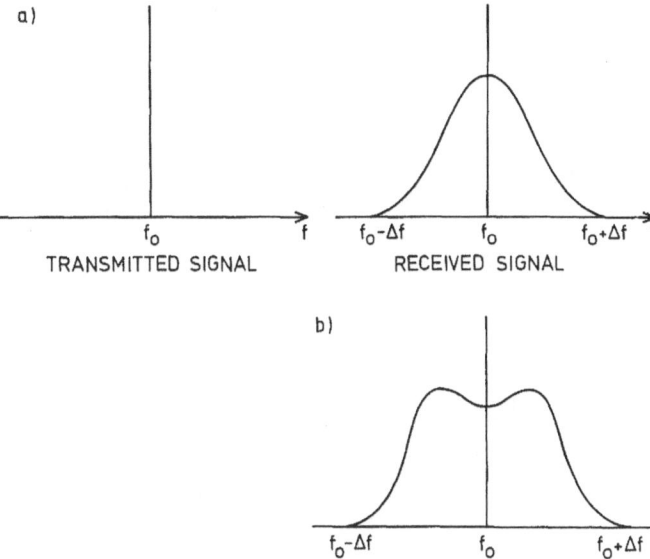

amount of positive charges. The positive charges in the ionosphere are tied to the ions, oxygen, nitrogen, nitric oxide, hydrogen etc. Therefore the movement of the electrons will, to some extent, also be influenced by the movement of the ions present in the ionosphere. If the ions and the electrons have the same temperature, and on the average are at rest with respect to the receiver, the scattered signal will have a bell-like spectrum centered at the transmitting frequency (Fig. 11.19). If, however, the electron temperature is higher than the ion temperature, the top of the bell will be suppressed and a double humped spectrum is formed. The width of the spectrum gives information on the mass and temperaturen of the ions and the temperature of the electrons and the area of the spectrum is a measure of the electron density at the height of scattering.

When the ions and electrons are moving together, as when a wind is present in the ionosphere, the spectrum will not be symmetric around the transmitted frequency, but around a frequency which is displaced according to the common velocity. It is, however, impossible to measure the true velocity of this wind from one receiver only as an infinite number of possible velocity directions exist which can give the same component along the line of sight of the receiver antenna. The EISCAT system (see coloured photograph on p. 123 and Fig. 11.17 and Fig. 11.20) operating at the highest frequency (1,290 MHz) has, in addition to the radar system at Ramfjordmoen, receiver stations at two other places, Kiruna, Sweden and Sodankylä, Finland. With this system it will be possible to observe components of the wind velocity vectors along three different directions, and thereby derive almost full information of the complete velocity vector of the auroral ionosphere.

With this system, which is one of the most advanced ground-based ionospheric research facilities in the world, it is possible simultaneously to derive fundamental parameters for an understanding of the structure, composition, and dynamics of the auroral ionosphere. This system therefore yields vital information for a better understanding of the processes taking place in the upper atmosphere during displays of northern lights.

In order to better utilize the EISCAT system, additional radio installations have been installed close to the EISCAT site. Almost 600 m × 600 m is now covered by radio antennas at Ramfjordmoen, which together with optical and geomagnetic installations, makes it the largest field site for auroral research in the world.

Fig. 11.20. A panoramic view of the gigantic installations at the Ramfjordmoen field site. (Photo K. Folkestad)

11.9 Andøya Rocket Range

The Andøya Rocket Range (Fig. 11.17 and Fig. 11.21) is located at the northern end of the island of Andøya in Northern Norway (69°18′N, 16°01′E) and can easily be reached by air, sea, and land. The island of Andøya is roughly 120 km southwest of Tromsø and 1,100 km north of Oslo.

The first rocket, a Nike/Cajun, was launched from the range on August 18, 1962. The range facilities at that time were kept to an absolute minimum, but it has since been developed into an important center for auroral research. At the end of 1981 a total of 202 instrumented rockets had been launched, including 30 different types of rocket configurations.

In addition to the rockets, a total of 158 scientific balloons ranging from 2,100 m³ to 50,000 m³ have been launched from the range. Scientific balloons

Fig. 11.21. The Andøya Rocket Range is located at the northern end of the island of Andøya (one of the outermost islands of the Vesterålen group) in Northern Norway. Southward the range is well protected by wild, impressive mountains. From the west to the northeast it faces the Norwegian Sea. This location, with a large sea impact area offers wide latitude in selecting rocket trajectories and permits launchings at an impact distance of more than 1,000 km. Thus, the rockets traverse the physically most interesting altitudes for auroral research

can carry instruments weighing almost 100 kg up to approximately 35 km altitude and drift at this level for a couple of days. The telemetry of data from the balloons is handled by the range.

Personnel from more than 70 research centres and universities in Europe, the U.S.A. and Japan have been engaged in scientific programmes carried out at the range. This clearly shows the attractiveness of "Norway's Cape Kennedy in miniature". Foreign scientists appreciate both the technical facilities at

151

the range and the local environment. The most significant factor, however, is Andøya's location relative to the auroral oval. The geomagnetic co-ordinates of the range are (67°15′ N and 114°23′ E) in the zone where there is maximum auroral activity during night time. The range is therefore ideal for studies of most phenomena in the polar atmosphere.

The Andøya Rocket Range has a large sea impact area permitting up to three-stage rockets to be launched more than 1,000 km away and allowing several choices of rocket trajectories which permit observations in different directions. The greatest peak altitude reached by a rocket so far has been approximately 800 km, but the altitude of most of the sounding rockets used in auroral studies does not exceed 300 km. This means, however, that the rockets traverse the physically most interesting altitude level for auroral research in near space. Total flight time of the rockets varies from 5–20 min, depending on the rocket configuration and the weight of the payload.

The control centre is the focal point of launch operations. Launch control, the telemetry station, and

Fig. 11.22. This DMSP photograph of northern Scandinavia, during quiet auroral conditions, shows the auroral oval crossing over the Andøya Rocket Range (Andenes). City lights outline the contours of Northern Europe. The location of the Range relative to, e.g., Bergen, and Tromsø is indicated in the photograph (see Fig. 10.12). (Courtesy Air Force Geophysical Laboratory)

Fig. 11.23. This picture shows two auroral payloads after the last check before launch. Notice that both rockets are filled with scientific instruments. Satellite and rocket instrumentation requires advanced technology. The instruments must work accurately in the thin plasma out in space

Fig. 11.24. The launch area with two rockets being prepared for take off. (At Andøya Rocket Range there are eight different launch pads). The vertical rocket is ready for take off, while the horizontal one is going through its last checks. The flight time of these auroral rockets varies between 5 and 10 min depending on the rocket configuration and the weight of the payload. During flight time several million discrete measurements of different auroral parameters are carried out

the experimenters' quick-look room are all housed here. In addition, the control centre contains such other facilities as workshops, offices, living rooms, and bedrooms.

Telemetry antennas located on the roof of the control centre and in adjacent areas track the rockets' flight towards the fascinating northern light, and all observation data are telemetered back and stored on magnetic tapes.

Various types of rockets may be launched from eight different launch pads in the launch area. To achieve the desired rocket trajectory, accurate information about the ground wind and high altitude winds is required. Continuous recordings of wind velocity and direction are obtained by anemometers placed 10, 25 and 50 m above the launch pads. Wind measurements, at up to a maximum altitude of 30 km, are made by tracking balloons carrying radar reflectors.

During countdown for which a visible northern light is a condition for launching, the optical site is the obvious place for the physicists involved. It is here that the project scientist has at his disposal a wealth of information from which to determine the right launching conditions. Readouts

from all ground-based instruments are available here, and from here he can see the night sky and be in direct contact with the payload manager and launch control.

If observations, from other ground-based stations in Northern Norway are necessary for determining the launch conditions, then communications and data links with these stations are easily established.

The rocket motors (usually two or three stages) constitute approximately 70% of the total rocket length which can be 10–15 m. Housed inside the nose cone and payload cylinder are usually 50–200 kg of complicated and ingenious instruments whose task it is to map the northern light through which the rocket flies. Even if the flight lasts only for a few minutes, several million observations are normally made by every rocket inside the auroral area.

The new and advanced installations for ground based studies of the upper atmosphere which are being established in northern Scandinavia will be key factors in any future scientific activity. The incoherent scatter radar, EISCAT, is expected to play a particularly important role. Ground-based facilities can monitor the state of the upper atmosphere and provide time series of observations, whereas detailed in

situ measurements are possible using rocket- and balloon-borne instruments. Co-ordinated "campaigns" using all these experimental tools will be important elements in future programmes. It may also be possible to co-ordinate satellite, ground based, and rocket techniques more effectively than in the past. New satellites are being planned with orbits which will make such co-ordinated studies feasible over northern Scandinavia.

The Royal Norwegian Council for Scientific and Industrial Research (NTNF), Space Activity Division, co-ordinates scientific and application research programmes in space activity which are performed at Norwegian universities, research institutes, and industrial plants. The Space Activity Division is responsible for the necessary infrastructure. An important part of this infrastructure is the Andøya Rocket Range.

12 Summary and Concluding Remarks

There are several naturally occurring heavenly phenomena which mankind is able to see and enjoy without the aid of anything more sophisticated than the eyes. The process of deduction from these observations, however, involves analyzing the information gathered by present, as well as past, observers and putting them into perspective. It is therefore not surprising that our views of these heavenly phenomena as well as of their causes have changed dramatically in the course of history.

Many bright objects appear in the sky on any clear, dark night. Of these the Moon, stars and the planets are the most dominating. Many of the stars appear in patterns or constellations as familiar objects to mankind. The ancients also noticed anomalous bright objects which moved very slowly among the night stars, and these they labelled "wanderers". Today, we know the "wanderers" are well-known planets.

One type of object in the sky which must have evoked wonder in the minds of shepherds and other night-watchers are the meteors – more generally known as "shooting stars". These meteors can appear sporadically, but they can also occur in the form of showers. Another spectacular light phenomenon, which may appear either in a sporadic or periodic manner is the comet. A comet can remain visible for several weeks, appearing before sunrise for a few days and then later after sunset for a period of time depending upon its direction of approach to the Sun. One of the most famous periodic comets was named after Edmund Halley (Chap. 5.2) who discovered it in 1682 and correctly predicted its return after 76 years. Since that time it has reappeared in 1758, 1834 and in 1910. We are now looking forward to its return in 1986.

Another interesting nighttime event is an eclipse of the Moon. Although not so spectacular as an eclipse of the Sun during the daytime, it is sufficiently unusual to create public interest even today. All phenomena described above occur over a large area of the Earth and in particular over those areas which are most densely populated.

Auroras (the northern and southern light), on the other hand, ordinarily appear in the so-called auroral zones which are two very remote areas of the Earth. Consequently, auroras are not readily available for viewing by the majority of people. Even today, the area of the southern auroral zone (roughly around Antarctica) is inhabited only intermittently. The northern auroral zone which crosses Alaska, Northern Canada, Northern Scandinavia, and Siberia was always accessible to the "nature" people (hunters and fishermen) living in the polar region, and with technical developments in this century, the area under the northern light has become permanently populated, even though sparsely.

It is therefore not surprising that historical records pertaining to northern lights and even early research reports are less numerous than in other fields of scientific research of naturally occurring heavenly light phenomena.

When we search the old (i.e. more than 1,000 years old) written records of northern light, we find the fountainhead of these efforts centered in the Mediterranean countries. Here, the great schools of learning and the libraries were established and flourished several centuries before the birth of Christ. On the other hand, an auroral display is seen in the Mediterranean area only after some unusually large solar activity. Thus, the time lapse between such large auroral events could easily be up to 50 or even 100 years. In addition, northern lights seen at such low latitudes are significantly less dramatic and colourful than those seen from the auroral zones but they were nevertheless strong enough to frighten people to promise to behave and better their ways.

During the Viking period (i.e. 500–1,300 A.D.) the development in knowledge of culture, arts, and science really started in the Scandinavian countries. This was to a large extent due to developing communications and commerce during this period. As we have pointed out in the text, a northern light can be seen practically every clear, dark night in the auroral zone. Therefore, we would expect the reaction to an auroral display in Northern Norway to be radically different from that in Greece. By comparing the description of the northern light in *The King's Mirror* and in old Roman and Greek literature, significant differences are found. Unfortunately, it appears that the northern auroral zone during the Viking period passed mainly over uninhabited areas, except for Greenland. Thus, only a small percent of the Vikings were familiar with this brilliant light phenomenon.

In this book we have mainly attempted to trace the contributions of the Scandinavians in the development of auroral research, from the beginning of scientific inquiry to the middle of our century. (The history of the aurora has been reviewed in several books and articles, and we will not repeat well-documented facts.)

We began our search through Norse mythological records. These are written in a language and style that are understood today only by a limited number of people. In the majority of translations of these old records, the emphasis has been directed more toward the literary than the scientific meaning. In fact, the original intent of the interpreters was directed more toward producing a work of literary art than a scientific treatise.

Nevertheless, we glean from these writings that the inhabitants of Greenland and the Nordic countries looked upon auroral displays in much the same manner as people living elsewhere reacted to the occurrence of meteors, comets, eclipses, etc. Therefore, to them the northern lights were a message from the gods of coming disasters, an association with deceased relatives, a battle of the gods, a weather sign, etc.

The first Norse literary work with a detailed realistic description of the northern light is *The King's Mirror* from ~1250 A.D. Even though the section devoted to northern lights is only a minute part of the famous chronicle, it is a very useful document. The appropriate name "Northern Light" is mentioned for the first time, and some reasonable speculations relative to the source of the phenomenon are included. Following *The King's Mirror*, no more written auroral records are found until about 1500 A.D. This is not too surprising because the period 1350 to 1500 A.D. was a poorly documented epoch.

From 1500 A.D. and particularly from the 18th and 19th century we have found several new records, directly or indirectly related to the northern light. The great majority of these sources were written by religious priests, from ministers to bishops. This is not too surprising since these were the only scholars during this period of time.

During this epoch there were lively speculations and differences of opinion as to the cause of the northern light, its height, and geographic location. All observations were made visually. Some beautiful drawings were done. The most important discovery was made in 1741 when Hiorter found that the northern light is related to disturbances in the geomagnetic field. By correspondence with other workers, it was then also concluded that both northern lights and geomagnetic disturbances are global in character, i.e. they occur simultaneously over fairly large areas. In the middle of the last century a correlation between sunspots and auroral occurrences was noted.

The First International Polar Year (1882/83) could be regarded as marking the beginning of modern auroral research, and the driving force behind the Scandinavian effort was Sophus Tromholt. His contributions set the stage for the brilliant work of Kristian Birkeland – the great auroral pioneer at the turn of the century. The later work on auroral problems by Størmer, Vegard, Harang, and Alfvén rests strongly on Birkeland's initial results and ideas.

Today, auroral research is mainly conducted through the use of sophisticated instruments on board rockets, satellites, and with advanced balloon and ground-based equipment. Many of the mysteries of the northern light have been solved partly or fully, but new problems have also appeared.

For those of us who admire the beauty of wonders in the dark sky, however, there is nothing comparable to a night with a magnificent auroral display. It is just as beautiful to watch today as it was in the early days of history. We know of no better poetic-description of a northern light than that of Tromholt who in his book *Under the Rays of the Aurora Borealis* from 1885 said:

"Lovely celestial display! Before your fascinating mysterious play, in which the enigmatic forces of Nature flood the heavens with light and colour throughout the long Polar night, the golden sunsets of the Pacific Ocean, the gorgeous flora of the Tropics, the resplendent lustre of the gems of Golconda, must pale.
Lovely celestial display!"

Appendix 1: List of Scandinavian Authors Who Have Contributed to the History of Northern Light Until 1800

Einar Gunnarson probably author of the "King's Mirror" about 1250 A.D.

Olaus Magnus (1490–1558) "Historia de Gentibus Septentrionalibus", Roma, 1550

Morten Pedersen (Alban) (1537–1594) Meteorological notes in "Paul Ebers Calendarium", Wittenberg, 1571

Absalon Pedersen Beyer (1528–1575) in his diary from 1552 to 1572

Tycho Brahe (1546–1601) in his Meteorological Journal

Peder Jacobsen Flemløse (ca. 1554–ca. 1598) „Elementiske og jordiske Astrologie om Luftens Forandring", Uraniborg, 1591

Peder Claussøn Friis (1545–1614) in "Om Grønland", Copenhagen, 1596 and 1604 or 1605

Christiernus Reitherus "Historico Geographica de Orbe Septentrional", Copenhagen, 1664

Suno Arnelius (1681–1740) "Exercitium Philosophicum de Chasmatibus", Uppsala, 1704

Thormod Thorfaeus (1636–1719) "Det Gamle Grønland", Copenhagen, 1706

Ole Rømer (1640–1710) „Descriptio – Luminis Borealis quod nocte inter 1 & 2 Febr. 1707 Hafniæ visum est", Miscellaneis Berolinensibus pp. 131–133, 1710

Jonas Ramus (1649–1718) „Norriges Beskrivelse", Copenhagen, 1715

Jens Christian Spidberg (1684–1762) „Historische Demonstration und Anmerkung über die Eigenschaften und Ursachen des sogenandten Nordlichts", Halle, 1724

A note in Nye Tidender om laerde og curieuse Sager, Copenhagen, 1738

Anders Celsius (1701–1744) „CCCXVI observationes de Lumine Boreali, ab a. MDCCXVI at a. MDCCXXXII par tim a de, par tim ab aliis in Svecia habitas", Norimbergæ, 1733

Johan Heitman (1664–1749) "Physiske Betænkninger over Solens Varme, Luftens skarpe Kuld og Nordlyset", Copenhagen, 1741

Peter Møller "Betænkninger over Nordlyset" and "Advertissement", Trondheim, 1741

Hans Egede (1686–1758) „Det gamle Grønlands Nye Perlustration eller Naturel Historie", Copenhagen, 1741

Samuel Triewald (1688–1742) „Experimentum Auroræ borealis Artificalis", Proc. Royal Swedish Acad. Sci., Vol. V, pp. 115–117, 1744

Joachim Frederik Ramus (1685 or 86–1769) „Historisk og Physisk Beskrivelse af Nordlysets forunderlige Skikkelse, Natur og Oprindelse", Proc. Acad. Sci. Copenhagen, Vol. I, pp. 317–396, 1743 and 1744; and Vol. III, pp. 148–212, 1747

Olof Peter Hiorter (1696–1750) „Om Magnet-Nålens Åtskillige ändringar", Proc. Royal Swedish Acad. Sci., pp. 27–43, 1747

Pehr Wilhelm Wargentin (1717–1783) "Observationer På Magnet-Nålen", Proc. Royal Swedish Acad. Sci., Vol. XI, pp. 52–59, 1750
"Vetenskapernas Historia om Norrskenet, Proc. Royal Swedish Acad. Sci., pp. 161–171, 1752
"Fortsättning af Historien om Norr-Skenet", Proc. Royal Swedish Acad. Sci., pp. 81–93, 1753

Lars Barhow (1707–1754) „Richtig angestellte und aufrichtig mitgeteilte Observationes von Phaenomeno Nordlicht", Leipzig, 1751

Erich Pontoppidan (1698–1764) „Det første Forsøg paa Norges Naturlige Historie", Copenhagen 1751 and 1753

Pehr Kalm (1716–1779) „Några Nordsken, observerade i Norra America", Proc. Royal Swedish Acad. Sci., pp. 145–155, 1752

Carl Gustaf Ekeberg (1716–1784) „Berättelse om et Norrsken, som observerades på utresan med Ostindiska Compagniets. Skepp, Sophia Albertina den 30. Jan. 1755", Proc. Royal Swedish Acad. Sci., pp. 61–64, 1757
„Observationer på Magnet-Nålens Inclination, Gjorda under Resan til och från Canton Årene 1766 och 1767", Proc. Royal Swedish Acad. Sci., pp. 225–228, 1768

Anders Wijkström (1726–1763) „Om Nordsken“, Calmar, 1759

Gerhard Schønning (1722–1789) „Nordlysets Ælde Beviist med gamle Skribenters Vidnesbyrd“, Proc. Acad. Sci. Copenhagen, pp. 197–316, 1759–1760 „Udtog af hvad der angaaede Vejrets af Luftens Forandringer og Beskaffenhed er iagttaget i Trondheim fra den 1ste Octob. 1759 til den 1ste Octob. 1761“, Proc. Acad. Sci. Copenhagen, Vol. 9, pp. 596–610, 1761–1764

Erich Johan Jessen (1705–1783) „Det Kongerige Norge fremstillet efter dets naturlige og borgerlige Tilstand“, Tom I, pp. 375–469, Copenhagen, 1763

Fredric Mallet (1728–1797) „Nordsken, observerad i Upsala den 17 October 1763“, Proc. Swedish Royal Acad. Sci., pp. 62–67, 1764

Torbern Olof Bergman (1735–1784) "Afhandling om Nordskenets högd", Proc. Royal Swedish Acad. Sci., pp. 193–211, and pp. 249–261, 1764

Johann Carl Wilcke (1732–1796) „Forsøk Til en Magnetisk Inclinations-Charta“, Proc. Royal Swedish Acad. Sci., pp. 193–225; 1768; „Om Magnet-Nålens årliga och dageliga ändringar i Stockholm“, Proc. Royal Swedish Acad. Sci., pp. 173–301, 1777

Eggert Olafsen and Bjarne Povelsen (1726–1768) „Reise igiennem Island“, Vol. II, pp. 907–910, Sorøe, 1771

H. Strøm (1726–1797) „Udtag af 12 Aars Meteorologiske Iagttagelser paa Søndmøre i Bergens Stift i Norge“, Proc. Royal Acad. Sci., Vol. XI, pp. 399–420, Copenhagen, 1777

Anders Hellant (1717–1789) „Magnet-Nålens Declination, observerad på flera Ställen inom Norra Pol-Circelen“, Proc. Royal Swedish Acad. Sci., pp. 300–305, 1777

Diderich Christian Fester (1732–1811) „Mathematiske og Physiske Betænkninger over Nordlyset. Første Betragtning“, Trondhjem, 1781 „Fortsatte Mathematiske og Physiske Betænkninger over Nordlyset“, Proc. Royal Norwegian Acad. Sci., Vol. II, pp. 124–182, 1788

Hans Arentz (1731–1793) „Beskrivelse over Søndfjord i det Nordre Bergenhusiske Amt“, Norges Topografiske Journal, XXIX, 24, 1785

Christian Ulrich Detlev Eggers (1758–1813) „Physikalische und statistische Beschreibung von Island aus authentischen Quellen und nach den neuesten Nachrichten“, Copenhagen, 1786

Christian Carl Lous (1724–1804) „Om Missvisningens her i Ciøbenhavn befundne Forandring i de sidste forløbne 50 Aar“, Proc. Royal Danish Acad. Sci., Vol. III, pp. 161–169, 1788

Thomas Bugge (1740–1815) „Om Nordlysets Indflytelse paa Magnetnaalens Declination iagttaget ved Gothaab i Grønland Aar 1786 til 1787 af Missionæren Herr Andreas Ginge“, Proc. Royal Danish Acad. Sci., Vol. III, pp. 531–549, 1788

Appendix 2: Map of Scandinavia Showing Locations and Areas Mentioned in the Text

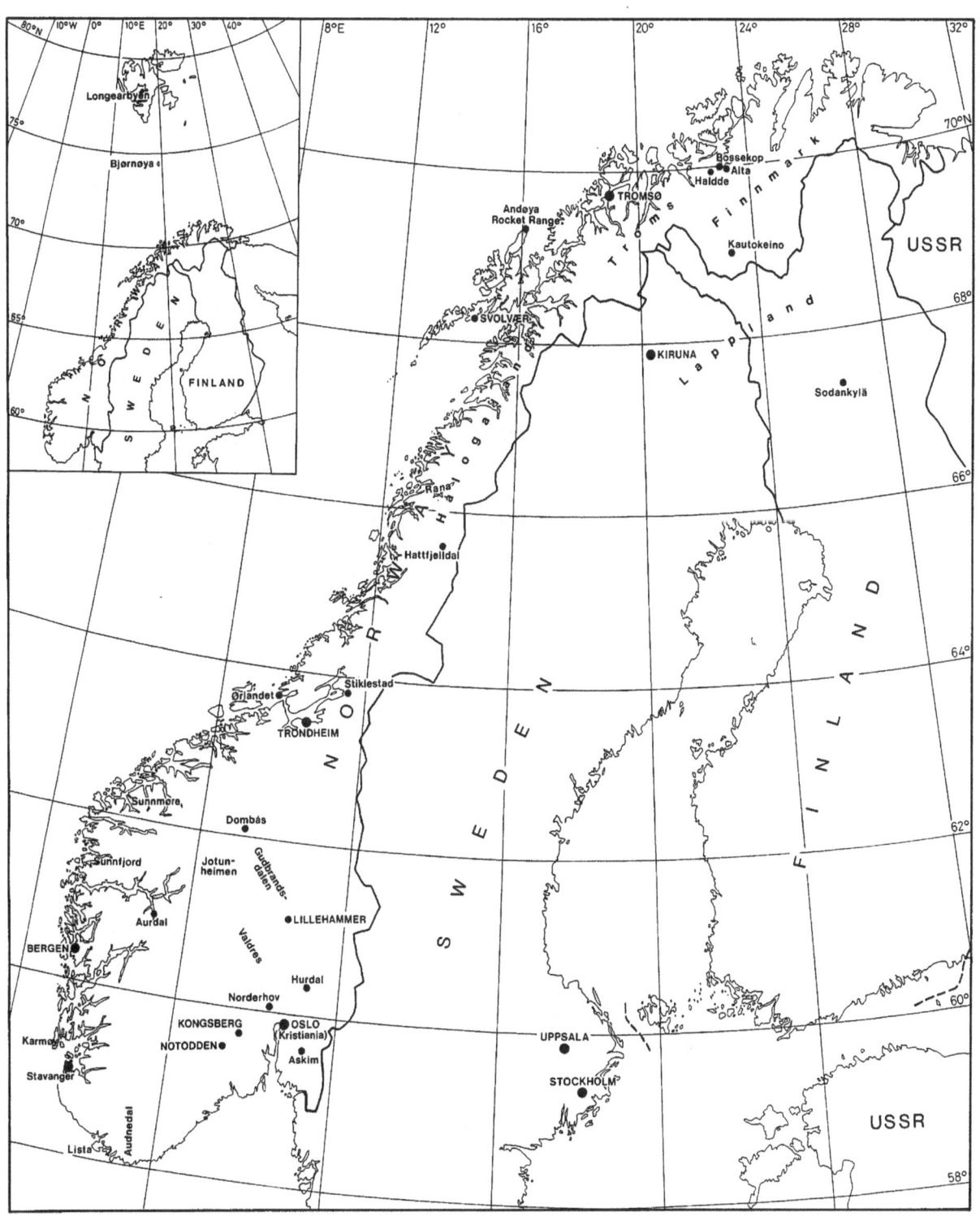

Appendix 3: Auroral Observatories

Some of the best known Observatories in the Northern Hemisphere

	Geographic		Geomagnetic	
	Latitude	Longitude	Latitude	Longitude
Ny-Ålesund, Spitzbergen	78°54'N	11°54'E	75.4	131.3
Thule, Greenland	77°29'N	69°10'W	89.0	358.0
Bjørnøya, Spitzbergen	69°14'N	53°32'W	79.9	32.5
Leirvågen, Iceland	64°11'N	21°42'W	70.2	71.0
Fort Churchill, Canada	58°48'N	94°06'W	68.7	322.8
Trømsø, Norway	69°36'N	18°54'E	67.1	116.8
Kiruna, Sweden	67°50'N	20°25'E	65.3	115.6
College, Alaska	64°52'N	147°50'W	64.6	256.5
Sodankylä, Finland	67°22'N	26°38'E	63.8	120.0
Murmansk, USSR	68°15'N	33°04'E	63.0	125.8
Dixon Island, USSR	73°33'N	80°34'E	63.0	161.6
Dombås, Norway	62°04'N	9°07'E	62.3	100.1
Tixi Bay, USSR	71°35'N	129°00'E	60.4	191.4
Lovö, Sweden	59°21'N	17°50'E	58.1	105.8
Nurmijärvi, Finland	60°31'N	24°39'E	57.9	112.6
Rude Skov, Denmark	55°51'N	12°51'E	55.9	98.5
Magnetic North Pole, ca.	76°N	102°W		
Magnetic South Pole, ca.	68°S	145°E		
Geomagnetic North Pole, ca.	78°31'N	70°W	90.0	

References

Arentz H (1785) Beskrivelse over Søndfjord i det Nordre Bergenhusiske Amt. Nor Topografisk J XXIX, 24, pp 21–26

Arnelius S (1704) Exercitium Philosophicum de Chasmatibus. Upsalis

Barhow L (1751) Richtig angestellte und aufrichtig mitgetheilte Observationes von Phaenomeno Nordlicht. Frankfurt und Leipzig

Bergman T (1764) Afhandling Om Nordskenens Høgd. Kongl. Sven Vetenskaps. Acad Handlingar, pp 193–210, pp 249–261

Birkeland K (1908, 1913) The Norwegian Aurora Polaris Expedition, 1902–1903. Aschehoug (Nygaard), Christiania, Vol. I, Vol II

Bjørgo T (1954) Mørker og Morgon, Dikt. Gyldendal Nor, Oslo

Braun GC (1819) Die Religion der alten Deutschen. Mainz

Bruheim JM (1977) Brevet til kjærleiken. Dikt. Nor Boklag, Oslo

Bugge S (1871) Tillæggsbemærkning om navnet Hålogaland, Helgeland. Hist Tidsskr. Kristiania 1:136–140

Caspari T (1921) Tidsskifte, Nyere Digte Aschehoug (Nygaard), Kristiania

Celsius A (1733) CCCXVI observationes de Lumine Boreali, ab a. MDCCXVI at a. MDCCXXXII par tim a de, par tim ab aliis in Svecia habitas. Norumbergæ

Dass P (1974) Norlands Trompet, udgitt ved Didrik Arup Seip. Aschehoug (Nygaard), Oslo

de Mairan JJ (1733) Traité Physique et Historique de l'Aurore Boréale. 1 st ed L'Imprimerie Royal, Paris

Eather RH (1980) Majestic Lights. Am Geophys Union, Washington DC

Eggers CUD (1786) Physikalische und Statistische Beschreibung von Island aus authentischen Quellen und nach den neuesten Nachrichten. Kjøbenhavn

Eskeland L (1921) Uppveg, Dikt. Bokreidar Lunde, Bjørgvin

Fester DC (1781) Matematiske og Physiske Betænkninger over Nordlyset Første Betraktning. Trondhjem

Fester DC (1788) Fortsatte Mathematiske og Physiske Betænkninger over Nordlyset. Det Kongl Nor Vidensk Selskabs Skrifter, Bd II, pp 124–182

Flemløse PJ (1591) Elementiske og jordiske Astrologie om Luftens Forandring. Uraniborg

Fritz H (1881) Das Polarlicht. Brockhaus, Leipzig

Halley E (1716) An account of the late surprizing appearance of the lights seen in the air, on the sixth of March last, with an attempt to explain the principal phenomena thereof. Phil. Trans. for the months of Jan, Feb, March. London

Hansteen C (1825) On the Aurora Borealis and Polar Fogs. Edinb Philos J Vol 12, pp 83–93

Hamsun K (1964) Samlede Verker, Bind 15. Gyldendal Nor, Oslo

Hardie EM (1883) The Midnight Cry; Behold the Bridegroom Cometh. Partridge, London

Harang L (1951) The Aurorae. Wiley, New York

Heitmann J (1741) Physiske Betænkninger over Solens Varme, Luftens skarpe Kuld og Nordlyset. Kjøbenhavn

Hell M (1770) Aurorae Borealis Theoria Nova, Pars I. Vindobonæ

Hellevik A (1965) Kongsspegelen. Det Nor, Samlaget

Hiorter OP (1747) Om Magnet-nålens Åtskillige ändringar. Kongl Swen Wetenskaps Acad Handlingar, pp 27–43

Hollander LM (1969) The Potic Edda. University of Dallas Press, Austin, Texas

Holm-Olsen L (1975) Edda, Dikt. Cappelens, Oslo

Holtsmark A (1964) Studier i Snorres Mytologi. Universitetsforlagte, Oslo

Hughes DW (1977) The inconstant Sun. Nature, Vol 266, pp 405–406

Jessen Schardebøll EJ (1763) Det Kongerige Norge fremstillet efter deres naturlige og borgerlige Tilstand. Tom I, pp 375–469, Kjøbenhavn

Kalm P (1752) Några Nordsken, observerade i Norra-America. Kongl Sven Acad Handlingar pp 145–155

Kawai N, Hiroka K (1967) Wobbing motion of the geomagnetic dipole field in historic time during these 2000 years. J Geom Geoel, Vol 19, pp 217–227

King JW, Hurst E, Slater AJ, Smith PA, Tamkin B (1974) Agriculture and sunspots. Nature 252: pp 2–3

Koht H (1920) Om namne Hålogaland. Haaløygminne, pp 3–11

Lanzerotti LJ (1979) Geomagnetic influence on manmade systems. J Atmos Terr Phys Vol 41, pp 787–796

Leem K (1767) Beskrivelse over Finnmarkens Lapper. Kjøbenhavn

Loomis E (1873) Comparison of the mean daily range of the magnetic declination and the number of auroras observed each year, with the extent of the black spots on the surface of the sun. Am J Sc Arts, Third Ser Vol V, No 28, pp 245–260

Magnus O (1976) Historia om De Nordiska Folken. Institut för folklivsforskning vid Nordiska museet och Stockholms Universitet

Magnusson F (1822) Den Ældre Edda. Kjøbenhavn

Magnusson F (1826) Æddalæren og dens Oprindelse, Kjøbenhavn

McDonald KL, Gunst RH An analysis of the earth's magnetic field from 1835 to 1965. ESSA Tech Rep, IER – 46 – IES – 1

Mortensson-Egnund I (1944) Edda – Kvæde. Norrøne Bokverk, Det Nor Samlaget, Oslo

Munch PA (1926) Norse Mythology. Oxford University Clarendon Press

Munch PA (1967) Nordens Gamle Gude- og Heltesagn. Universitetsforlaget, Oslo

Møller P (1741) Betænkninger over nordlyset, Advertissement. Trondheim

Nansen F (1898) Farthest North. Newnes, London

Nordenskiöld AE (1882–83) Rapporter skrivna under loppet af Vegas Expedition. Om Norskenen. Stockholm

Olsen M (1909) Fra Gammelnorsk Myte og Kultus. Maal og Minne, Oslo

Petrie W (1963) Keoeit, The Story of the Aurora Borealis. Pergamon, New York

Pontoppidan E (1752) Det første Forsøg paa Norges Naturlige Historie. Kjøbenhavn

Ramus J (1715) Noriges Beskrivelse. Kjøbenhavn

Ramus JF (1747) Historisk og Physisk Beskrivelse av Nordlysets forunderlige Skikkelse, Natur og Oprindelse. Det Kiøbenhavnske Selskab, Skrifter, Vol. I, 1745, Tredie Deel

Reitherus C (1664) Historico Geographica de Orbe Septentrionale. Hafuiæ

Rode F (1842) Optegnelse fra Finnmarken. Skien

Rothly A, Berkes Z (1963) Nordlicht – Beobachtungen in Ungarn. Akad Kiado, Budapest

Rubenson R (1879, 1882) Catalogue des aurores boréales observées en Suède. Kungl Sven Vetenskap Acad Handlingar 15(5), 18(1)

Rømer O (OR) (1710) Descriptio – Luminis Borealis quod nocte inter 1 & 2 Febr. 1707. Hafniæ visum est. Misc Berolinensibus pp 131–133

Sande J (1969) Krossen og Sleggja, Dikt i utvalg. Gyldendal Nor, Oslo

Schøning G (1760) Nordlyset Ælde Bevist med gamle Skribenters Vidnesbyrd. Det Kiøbenhavnske Selskab. Skrifter Vol 8, pp 197–316

Smith RC (1976) Reise i Norge 1838, translated by Johnny Johnsen. Universitetsforlaget, Oslo Bergen Tromsø

Spidberg JC (1724) Historische Demonstration und Anmerkung über die Eigenschaften und Ursachen des sogenannten Nordlichts. Halle

Sturlasson S (1929) Den Norrøne Gudeheimen, Snorre – Edda. Nordlis, Oslo

Sturlasson S (1979) Noregs konge soger. Det Nor Samlaget, Oslo

Sverdrup HU (1949) Evig Byggende Babel, Dikt. Gylendal Nor, Oslo

Tacitus C (1969) Germania. Natur och Kultur, Stockholm

Trap Wahl V (1909) Løvetand, Dikte. Aschehoug (Nygaard), Kristinia

Triewald S (1744) Experimentum Aurora borealis Artificials. Kongl Sven Wetenskap Acad Handlingar, Vol V, pp 115–117

Tromholt S (1885) Under Nordlyset Straaler. Gyldendal Boghandels, Kjøbenhavn

Tromholt S (1902) Katalog der in Norwegen bis Juni 1878 beobachteten Nordlichter. Dybwad, Kristiania

Wargentin PW (1750) Observationer på Magnet-Nålen. Kongl Sven Acad Handlingar, Vol XI, pp 52–59

Wargentin PW (1752, 1753) Vetenskapernas Historia om Nordskenet. Kongl Sven Vetenskaps Acad Handlingar, pp 161–171, pp 81–93

Wilcke JC (1768) Forsøk Til en Magnetisk Inclinations – Charta. Kongl Sven Vetenskabs Acad Handlingar, pp 193–225

Ørsted HC (1826) Oversigt over Det Kongelige Danske Videnskabernes Selskabs Forhandlinger. Det Kongl Dan Vidensk Selskab Avhandl, Anden Deel, Kjøbenhavn

The following books are not specifically referred to in the text, nevertheless the authors have benefited heavily from them.

Alexander HB (1916) The Mythology of all Races. North American, Vol X, Jones, Boston

Alfvén H (1963) Cosmical Electrodynamics. Oxford Clarendon Press

Angot A (1897) The Aurora Borealis, Appleton, New York

Arrhenius S (1900) Über die Ursache der Nordlichter. Kongl Vetenskaps-Akad Förhandlingar, No 5, pp 545–580

Arrhenius S (1904) On the Electric Equilibrium of the Sun. Proc R Soc Lond Vol 73, pp 496–499

Bell Mackintosh J (1903) The Fireside Stories of the Chippwyans. J Am Folklore Vol XVI pp 73–84

Best E (1924) The Maori. Mem Polynesian Soc Vol V, p 217, Wellington NZ

Birkeland H (1954) Nordens Historie i Middelalderen etter Arabiske Kilder. Skrifter Nor Videnskaps-Akad. Oslo, II, Hist-Filos Klasse, No. 2

Birkeland K (1901) Expédition Norvègienne pour l'étude des aurores boréales. Resultats des recherches magnetique. Christiania

Birkeland K (1913) Om verdens tilblivelse. Festskrift for Aars og Voss skoles femti års jubileum. pp 214–246, Kristiania

Birkeli E (1944) Huskult og Hinsidighetstro Skrifter Nor Videnskaps-Akad. Oslo, II, Hist-Filos Klasse, No. 1

Brington D (1898) Religions of Primitive Peoples. American Lectures on the History of Religions, Putnam, New York

Brun V (1962) Regnekunsten i det gamle Norge. Universitetsforlaget, Oslo

Devik O (1919) Kristian Birkeland. Teknisk Ukeblad, Nr 40, pp 969–972

Devik O (1968) Kristian Birkeland as I knew him, in the Birkeland Symp on aurora and Magnetic Storms. p 13 eds Egeland A and Holtet J. Centre National de la Recherche Scientific, Paris

Daae L (1888) Italieneren Francesco Negris Reise i Norge 1664-1665. Hist Tidsskrift, Bind 6, pp 85–158, Kristiania

Eddy JA (1976) The Maunder Minimum. Science 192, pp 1189–1202

Egede HP (1722) Relation angaaende dend Dessein med dend Grønlandske Mission. pp 63, Kjøbenhavn

Egede H (1741) Det gamle Grønlands Nye Perlustration eller Naturel Historie. Kjøbenhavn

Egeland A, Holter Ø, Omholt A (1973) Cosmical Geophysics. Universitetsforlaget, Oslo

Egeland A, Omholt A (1966) Carl Størmers height measurements of aurora. Geophys Publ 26 (6) 1

Einbu S (1945) Måne- og stjernekultus nørdst i grendom. Årbok for Gudbrandsdalen, pp 70–74, Dølaringen Boklag, Lillehammer

Encyclopaedia of Superstitions, Folklore and the Occult Sciences of the World, Vol II, pp 935–936, Detroit, 1971

Fearnley F (1859) Bestemmelse af Nordlysets Høide ved Målinger af en Iagttager. Kristiania Videnskabs Selskabets Handlinger, Bind 8, pp 117–140

Freuchen PB (1907) Om Nordlys. Fysisk Tidsskr, pp 89–95, Kjøbenhavn

Friis PC (1881) Samlede Skrifter. Udgivne for den Nor Hist Forening af Dr Gustav Storm. Kristiania

Hamilton JC (1903) The Algonquin Manabozho and Hiawatha. J Am Folklore, 16 pp 229–233

Hansteen C (1827) On the polar lights, or aurora borealis and australis. Phil Mag 2, 234

Hermundstad K (1940) Bondeliv, Gamal Valdres-Kultur II. Nor Folkeminnelag Nr 45, pp 127–130, Oslo

Hoffmann-Krayer E (1934–1935) Handwörterbuch des Deutschen Aberglaubens. Band VI, pp 1118–1121. de Gruyter, Berlin Leipzig

Holberg L (1750) Epistler Tomus II, IV og V. Kjøbenhavn

Holmberg U (1927) The Mythology of all races. Finno Ugric Siberian Vol IV, Archaeological Institute of America

Holzworth RH (1975) Folklore and the Aurora. EOS, Vol 56, No 10, pp 686–688

Haavio M (1944) Volkstümliche Auffassungen vom Nordlicht. Sitzungsber Finn Akad Wiss 1943, pp 199–226, Helsinki

Itkonen TI (1946) Heidnische Religion und späterer Aberglaube bei den Finnischen Lappen. Suom Ugrilainen Seura, Helsinki

Jobes G (1962) Dictionary of Mythology Folklore and Symbols. Part I, pp 158–159, Scarecrow, New York

Jones W (1911–12) Notes on the Fox Indians. J Am Folklore, 24, pp 209–237

Judson KB (1910) Myths and Legends of the Pacific Nortwest

La Cour O (1907) Adam Frederik Wivet Paulsen. Fysisk Tidsskr Kjøbenhavn

Leach M (ed) Standard Dictionary of Folklore Mythology and Legend. Volume A-1, p 91. Funk and Wagnalls New York

Lemstrøm KS (1869) Observationer på luftelektriciteten och polarljuset under 1868 års Svenske polarexpedition, Kongl Sven Vetenskaps Akad Förhandlingar No 7, pp 663–689

Lemstrøm KS (1873) Om den elektriska urladdningen i Polarljuset och Polarljus-Sepctrum. Frenckell, Helsingfors

Lemstrøm KS (1886) Om Polarljuset eller Norrskenet. Bonnier, Stockholm

Liestøl A (1971) Njålsoga. Det Nor Samlaget, Oslo

Lundmark B (1976) Det hörbara ljuset. Västerbotten, No 1/2 pp 86–89

Mackenzie DA (1935) Scottisch Folk-Lore and Folk Life. pp 85–98, pp 221–223, Blackie, London Glasgow

Mallet F (1764) Nordsken, observeradt i Upsala den 17 October 1763. Kongl Sven Videnskaps Acad Handlingar, pp 63–67

Nansen F (1961) På ski over Grønland Eskimoliv. Aschehoug (Nygaard) pp 191–215, Oslo

Nelson RK (1969) Hunters of the Northern Ice. p 138, The Univ Chicago Press

Nordgaard O (1912) Folkemeteorologi eller Gamle Merker for Veir og Vekst. Det Kongl Nor Videnskabers Selskabs Skrifter, Nr 8 pp 1–43

Nordland O (1968) Folklore and religion among the northern people – a contribution to the discussion of the Arctic Circumpolar Theory. Proc 8th Int Conf Anthropol Ethnol Sci 2, 305–310

Omholt A (1971) The Optical Aurora. Springer, Berlin Heidelberg New York

Opedal HO (1934) Makter og Menneske, Folkeminne ifrå Hardanger. Nor Folkeminnelag, Nr 32, pp 30–36, Oslo

Paulsen A (1894) Sur la nature et l'origine de l'aurore boréale. Det Kongl Dan Videnskabers Selskabs Förhandlingar, pp 148–168

Paulsen A (1895) Effet de l'humidité de l'air et action du champ magnétique terrestre sur l'aspect de l'aurore boréale, Det Kongl Dan Videnskabers Selskabs Förhandlinger pp 279–302

Paulsen A (1896) Nordlysets Straalingsteori, Nyt-Tidsskrift for Fysik og Kemi. Første Bind, Det Nordiske, pp 161–172

Paulsen A (1897–98) Nordlyset. Nord og Syd pp 689–711, Kjøbenhavn

Paulsen A (1906) Sur les récentes Théories de l'Aurorae Polaire. Det Kongl Dan Videnskabers Selskabs Forhandlinger, No 2, pp 1–37

Radford E, NA (1969) Encyclopedia of Superstitions. Ed Hole Ch, p 26, Hutchinson, London

Rasmussen K (1921–24) Thule-Expedition 5, Vol 7:I p 95, Kjøbenhavn

Rasmussen K (1978) Oppfattelse af Naturen Myter og Sagn fra Grønland (1921). Nordiske Lands Bogforlag

Ratcliffe JA (1970) Sun, Earth and Radio. An Introduction to the Ionosphere and Magnetosphere. World University Library, McGraw-Hill, New York

Ray OJ (1958) Legends of the Northern Lights. The Alaska Sportsman, pp 20

Røstad A (1933) Gamle tankar og truer um Nordljoset. Syn of Segn p 166–177

Spencer R (1976) The North Alaskan Eskimo. A Study in Ecology and Society. pp 258–259, Dover New York

Storaker JTh (1923) Rummet i Den Norske Folketro, Nor Folkeminnelag VIII, Kristiania

Storaker JTh (1924) Elementene i Den Norske Folketro. Nor Folkeminnelag X, Kristiania

Størmer C (1911) Bericht über eine Expedition nach Bossekop zwecks photographischer Aufnahmen und Höhenmessungen von Nordlichtern. Videnskaps Selskapets Skrifter, Math Nat Klasse, Kristiania

Størmer C (1955) The Polar Aurora. Oxford Clarendon Press

Sundell AF (1905) Minnestal öfver Karl Selim Lemström. Fin Vetenskaps Soc Förhandlingar, pp 1–22

Sæland S (1918) Professor Kristian Birkeland. Fysisk Tidsskr, pp 34–53, Kjøbenhavn

Torfessøn T (Thorfæus) (1927) Det gamle Grønland eller Det gamle Grønlands Beskrivelse. Oslo Etnografisk museum, Oslo

Tromholt S (1882, 1883) Om Nordlyset. Naturen No 6, pp 81–87, No 7, pp 97–104, No 8, pp 113–118, No 9, pp 168–172, 1882. No 6, pp 81–89, 1883

Tromholt S (1902) Katalog der in Norwegen bis Juni 1878 beobachteten Nordlichter. Oslo

Turner LM (1984) Ethnology of the Ungava District, Hudson Bay Territory. Bur Ethnol Secretary of the Smithsonian Inst, Government Printing Office, Washington

Vegard L (1916) Nordlichtuntersuchungen. Videnskaps Selskapets. Skrifter, Math Nat Klasse, Kristiania

Vegard L (1955) Kristian Birkeland. Nor Hydro gjennom 50 år, pp 579–585, Ed. K. Anker Olsen, Oslo

von Urbanitzky AR (1887) Elektrizität und Magnetismus im Alterthume. pp 110–127, Wiesbaden

Werlauff EC (1836) Bidrag til den Nordiske Ravhandels Historie. Det Kongl Dan Videnskabers Selskabs Avhandlinger, Vol 5, Kjøbenhavn

Wijkström A (1759) Om Nordsken. Calmar

Wilcke JC (1777) Rön om Magnet-Nålens årliga och dagliga ändringar i Stockholm. Kongl Sven Vetenskaps Akad Handlingar, pp 273–300

Qvigstad J (1927) Lappiske Eventyr og Sagn. Inst for Sammenlignende Kulturforskning, Aschehoug (Nygaard), Oslo

Ångström AJ (1968) Recherches sur le spectre Solaire, 4°, Upsala

Name Index

Page numbers in *italics* indicate portraits of authors

Subject Index

Physics and Chemistry in Space

Editors: J. G. Roederer, J. T. Wasson

Springer-Verlag
Berlin
Heidelberg
New York
Tokyo

Journal of Geophysics Zeitschrift für Geophysik

Edited for the Deutsche Geophysikalische Gesellschaft
by W. Dieminger, G. Müller, J. Untiedt

Editorial Board: K. M. Creer, Edinburgh; W. Dieminger,
Lindau üb. Northeim/Hannover; C. Kisslinger, Boulder, CO;
Th. Krey, Hannover; G. Müller, Frankfurt; G. C. Reid,
Boulder, CO; J. Untiedt, Münster/Westfalen; S. Uyeda, Tokyo
in collaboration with a distinguished advisory board.

The **Journal of Geophysics** publishes articles predominantly in
English from the entire field of geophysics, including original
essays, short reports, letters to the editor, book discussions, and
reviews articles of current interest, on the invitation of the German Geophysical Association. The following fields of geophysics have been treated in recent volumes: applied geophysics,
geomagnetism, gravity, hydrology, physics of the solid earth,
seismology, physics of the upper atmosphere, and volcanology.

Fields of Interest: Geophysics, Seismology, Geomagnetism,
Aeronomy, Extraterrestical Physics, Space Research, Meteorology, Oceanography, Applied Geophysics, Theoretical Geophysics, Tectonics, Geochemistry, Petrology.

 Reduced rate for members of the
Deutsche Geophysikalische Gesellschaft

**Springer-Verlag
Berlin
Heidelberg
New York
Tokyo**

Subscription information and/or **sample copies** are available
from your bookseller or directly from
Springer-Verlag, Journal Promotion Dept.,
P. O. Box 105 280, D-6900 Heidelberg, FRG